SCIENCE AND CHRISTIAN ETHICS

There is a growing crisis in scientific research characterized by failures to reproduce experimental results, fraud, lack of innovation, and burnout. In *Science and Christian Ethics*, Paul Scherz traces these problems to the drive by governments and business to make scientists into competitive entrepreneurs who use their research results to stimulate economic growth. The result is a competitive environment aimed at commodifying the world. In order to confront this problem of character, Scherz examines the alternative Aristotelian and Stoic models of reforming character, found in the works of Alasdair MacIntyre and Michel Foucault. Against many prominent virtue ethicists, he argues that what individual scientists need is a regime of spiritual exercises, such as those found in Stoicism as it was adopted by Christianity, in order to refocus on the good of truth in the face of institutional pressure. His book illuminates pressing issues in research ethics, moral education, and anthropology.

PAUL SCHERZ is Associate Professor of Moral Theology and Ethics at The Catholic University of America. He publishes broadly and teaches in the fields of bioethics and the relationship between religion and science.

NEW STUDIES IN CHRISTIAN ETHICS

Christian ethics has increasingly assumed a central place within academic theology. At the same time, the growing power and ambiguity of modern science and the rising dissatisfaction within the social sciences about claims to value neutrality have prompted renewed interest in ethics within the secular academic world. There is, therefore, a need for studies in Christian ethics which, as well as being concerned with the relevance of Christian ethics to the present- day secular debate, are well informed about parallel discussions in recent philosophy, science, or social science. New Studies in Christian Ethics aims to provide books that do this at the highest intellectual level and demonstrate that Christian ethics can make a distinctive contribution to this debate – either in moral substance or in terms of underlying moral justifications.

TITLES PUBLISHED IN THE SERIES

1. *Rights and Christian Ethics*, Kieran Cronin
2. *Biblical Interpretation and Christian Ethics*, Ian McDonald
3. *Power and Christian Ethics*, James Mackey
4. *Plurality and Christian Ethics*, Ian S. Markham
5. *Moral Action and Christian Ethics*, Jean Porter
6. *Responsibility and Christian Ethics*, William Schweiker
7. *Justice and Christian Ethics*, E. Clinton Gardner
8. *Feminism and Christian Ethics*, Susan Parsons
9. *Sex, Gender and Christian Ethics*, Lisa Sowle Cahill
10. *The Environment and Christian Ethics*, Michael Northcott
11. *Concepts of Person and Christian Ethics*, Stanley Rudman
12. *Priorities and Christian Ethics*, Garth Hallett
13. *Community, Liberalism and Christian Ethics*, David Fergusson

(*continued after the Index*)

SCIENCE AND CHRISTIAN ETHICS

PAUL SCHERZ
The Catholic University of America

CAMBRIDGE
UNIVERSITY PRESS

CAMBRIDGE
UNIVERSITY PRESS

University Printing House, Cambridge CB2 8BS, United Kingdom

One Liberty Plaza, 20th Floor, New York, NY 10006, USA

477 Williamstown Road, Port Melbourne, VIC 3207, Australia

314–321, 3rd Floor, Plot 3, Splendor Forum, Jasola District Centre, New Delhi – 110025, India

79 Anson Road, #06–04/06, Singapore 079906

Cambridge University Press is part of the University of Cambridge.

It furthers the University's mission by disseminating knowledge in the pursuit of education, learning, and research at the highest international levels of excellence.

www.cambridge.org
Information on this title: www.cambridge.org/9781108482202
DOI: 10.1017/9781108593694

First published 2019

Printed and bound in Great Britain by Clays Ltd, Elcograf S.p.A.

A catalogue record for this publication is available from the British Library.

Library of Congress Cataloging-in-Publication Data
NAMES: Scherz, Paul J., author.
TITLE: Science and Christian ethics / Paul Scherz, The Catholic University of America.
DESCRIPTION: First edition. | New York : Cambridge University Press, 2019. | Includes bibliographical references and index.
IDENTIFIERS: LCCN 2018054279| ISBN 9781108482202 (hardback) | ISBN 9781108742580 (pbk.)
SUBJECTS: LCSH: Science–Moral and ethical aspects. | Christian ethics. | Religion and science.
CLASSIFICATION: LCC BJ57 .S26 2019 | DDC 241–dc23
LC record available at https://lccn.loc.gov/2018054279

ISBN 978-1-108-48220-2 Hardback

Contents

General Editor's Preface

Two of the previous contributors have had considerable practical experience as working academic scientists before writing their monographs for *New Studies in Christian Ethics*. The first was Celia Deane-Drummond in *Genetics and Christian Ethics* (2006). Her original doctorate was in plant physiology; she held post-doctoral science fellowships at the Universities of British Columbia and Cambridge; and then became a Lecturer in plant physiology at Durham University. Only then did she turn to academic theology. The second was Christopher C. H. Cook in *Alcohol, Addiction and Christian Ethics* (2006). He originally qualified in medicine and trained as a psychiatrist, in time becoming Professor of the Psychiatry of Alcohol Misuse at the University of Kent. He too turned to academic theology at this midpoint in his career. Paul Scherz now joins them (as he explains in his Introduction) with a number of years of practical work as a laboratory biologist before his turn to academic theology. It is one thing for Christian ethicists without any significant background in practical science to write on science-related topics, but it is quite another for these three Christian ethicists. They bring a level of knowledge, experience, and credibility in scientific practice that the rest of us cannot quite match, admirably fulfilling the two central aims of *New Studies in Christian Ethics*:

1. To promote monographs in Christian ethics that engage centrally with the present secular moral debate at the highest possible intellectual level.
2. To encourage contributors to demonstrate that Christian ethics can make a distinctive contribution to this debate – either in moral substance or in terms of underlying moral justifications.

Paul Scherz's focus is on modern scientific culture, offering at the outset timely examples of how the entrepreneurial ideal is increasingly distorting scientific research – encouraging researchers to think of themselves as market actors and their work as being primarily focused on business opportunities. He is especially concerned with three aspects of what many have called a contemporary crisis in science: (1) problems in reproducibility, (2) a lack of innovation despite increasing hyperbole about the promise of results, and (3) increasing levels of burnout among researchers. With his own direct experience of laboratory research, he is well placed to make these critical points in the opening chapters of this new monograph and then, with his additional theological training, to address them ethically and theologically in subsequent chapters. In the latter it is evident that he has a particular and critical interest in the works of Michel Foucault and Alasdair MacIntyre. He argues that a combination of Aristotelian virtues (following MacIntyre) and Stoic moral techniques (following Foucault) is needed within scientific culture today if its problems are to be addressed adequately. Stoic moral formation in particular, so he and a few other theologians have argued recently, presupposes that virtuous practices need to be cultivated in the midst of communities and social contexts that are not virtuous themselves but are, in various ways and to various degrees, morally distorted. More specifically, Christian faith and practice can, so he believes, deepen the risky truth-speaking that is needed (and sometimes lacking) within the scientific community, by tying it to trust and faith in God.

There is much here to stimulate further discussion. Paul Scherz offers a well-informed, original, and significant challenge to current scientific culture that, I believe, deserves to be taken very seriously indeed.

Acknowledgments

The impetus for this work began long ago, during my scientific training, and required the support and encouragement of a host of people to come to fruition. My deepest debt of gratitude belongs to my mentors, Bill Skarnes, Cliff Tabin, and Didier Stainier, and all of my lab mates in their laboratories. While I could feel the pressures of entrepreneurial science growing during my education, I was lucky enough to be trained by supervisors and colleagues who gave me an example of the best forms of scientific life. It is this training that provided the experiences and resources that allowed for the critique of the problems of contemporary science found in this book.

Though extensively reworked to address the problems of contemporary science, much of the analysis in this book arose from insights in my dissertation research in the Theology Department at the University of Notre Dame. There I benefited from my fellow students, including Luis Vera, Jeff Morgan, David Elliot, Angela Carpenter, and Brian Hamilton, who served as interlocutors throughout my graduate work, especially in the form of the moral theology dissertation writing group. Phil Sloan gave important advice on sections of this project. Jean Porter and Gretchen Reydams-Schils helped me to develop my understanding of Aristotelian and Stoic ethical theory and ensured that that training was well applied in this project. The project would not have been possible without the criticism and encouragement of my advisor, Jerry McKenny. His interest in Foucault's late ethics gave me the push I needed to start this project, and he has continued to provide invaluable support throughout its development into a book. Jerry is simply the best advisor I could have hoped to have had.

My colleagues at the Catholic University of America have been extremely helpful in the development of this project. I would especially like to thank John Grabowski, David Cloutier, and Bill Mattison, who read sections of this book and gave me important advice. I thank the School, especially

Dean Mark Morozowich, for granting a teaching release in order to complete revisions on the book. Catholic University's excellent students have helped me to refine parts of this book, including members of a doctoral seminar on Social and Spiritual Practices. Beth Lofgren, Julie Amajuoyi, and Mariele Courtois served as research assistants for this project thanks to the generosity of the School. Mariele read and commented on the entire manuscript, catching many infelicitous phrases and confusing sentences.

This work benefited from the questions and comments I received following presentations on aspects of this project at Cambridge's Faculty of Divinity, Yale Divinity School, Marquette's Department of Theology, Notre Dame's Virtues working group and Science as Practice Group, Harvard Divinity School's Ways of Knowing Conference, the University of Virginia School of Engineering, and the Conference on Medicine and Religion. The Theology, Medicine, and Culture Initiative at Duke Divinity School was kind enough to host a manuscript workshop where I benefited from the comments and support of Farr Curlin, Stanley Hauerwas, Warren Kinghorn, Brett McCarty, Paul Griffiths, Gopal Sreenivasan, Richard Payne, and Jeffrey Baker. Jeff Bishop and other participants in conferences of the International Academy for Bioethical Inquiry have given me valuable feedback.

This book includes material adapted from prior publications. Sections of Chapters 1 and 2 were developed from my articles "Knowledge and the Scientist-Entrepreneur," in *Pro Ecclesia* 24 (2015): 175–92; and "Trivial Pursuits: The Decline of Scientific Research," in *The Hedgehog Review* 18, no. 3 (2016): 80–9. Part of Chapter 6 is adapted from "Legal Suppression of Scientific Data and the Christian Virtue of *Parrhesia*," in the *Journal of the Society of Christian Ethics* 35, no. 2 (2015): 175–92 (doi: 10.1353/sce.2015.0024). I thank the editors of these journals for the permission to use this material.

It has been wonderful to work with Cambridge University Press. I owe a debt of gratitude to Robin Gill, who has been an insightful, critical, and supportive editor throughout the entire process. Beatrice Rehl has been very helpful in guiding me through the publication process, as has Eilidh Burrett.

This intellectual support was essential to this project, but it also relied on material support from a number of sources. The generosity of Richard and Peggy Notebaert provided the fellowship that funded my graduate training in theology. The Issachar Fund assisted me in securing funding for developing the ideas from my dissertation into this book. The Institute for

Human Ecology at the Catholic University of America, at which I am a faculty fellow, provided research funding as well as a venue to present my work. The Institute for Advanced Studies in Culture at the University of Virginia, at which I am a visiting faculty fellow, generously provided me with a home away from campus, a venue to present my work, and a rich community of interlocutors, especially Joe Davis, James Hunter, Matt Crawford, and Jay Tolson.

Finally, I thank my family, who has supported me through the long process of completing and revising this work. My children, Iggy and Lucy, filled my weekends and vacations with joy, reminding me that a scholar is never only a scholar, but partakes of even more important vocations. Without China Scherz's encouragement to pursue these ideas and concerns by completing a second doctorate in a completely different field, this book would never have come to be. She has read every word of this book many times after having already discussed all of these ideas in embryonic form before they ever reached the page.

Introduction

There is a widespread recognition that there are significant problems in contemporary science. Social scientists, ethicists, and scientists themselves are concerned about problems in the reliability of scientific results, in the development of innovative insights, and in the declining quality of the working conditions of most researchers. It is the contention of this book that Christian ethics, through its dialogue with social science and philosophy, has resources to help individual researchers and thus the research enterprise as a whole to confront these problems in science. To provide these resources, however, Christian ethics must engage in a somewhat different discussion of the relationship between religion and science than is usually encountered in theological discussions of science.

Many valuable theological works have been written about Christianity and science: reactions to the claims of the New Atheists that defend the compatibility of theology and science, attempts to show the implications of scientific results for Christian ethics, and discussions of how to ethically use the knowledge resulting from scientific research.[1] Yet, reflecting on my own experience as a researcher, there seems to be a lacuna in this valuable literature. I spent many years in the laboratory: two years as an undergraduate researcher studying the genetics of early mouse development at Berkeley; three and a half years in doctoral research at Harvard on limb

[1] The literature in these areas is vast. For reactions to the New Atheists, see Peter Harrison, *The Territories of Science and Religion* (Chicago: University of Chicago Press, 2017); Alister E. McGrath, *Dawkins' God: From the Selfish Gene to the God Delusion* (Malden, MA: Wiley, 2015). For recent examples of theologians engaging the insights of science, see Martin A. Nowak and Sarah Coakley, eds., *Evolution, Games, and God: The Principle of Cooperation* (Cambridge, MA: Harvard University Press, 2013); Celia Deane-Drummond, *Christ and Evolution: Wonder and Wisdom* (Minneapolis, MN: Fortress Press, 2009); Stephen J. Pope, *Human Evolution and Christian Ethics* (New York: Cambridge University Press, 2007). For discussions of the ethics of using the results of science, see Gerald McKenny, *Biotechnology, Human Nature, and Christian Ethics* (New York: Cambridge University Press, 2018); Celia Deane-Drummond, *Genetics and Christian Ethics* (New York: Cambridge University Press, 2005).

development in the mouse and chick; and two years in a postdoctoral fellowship at UCSF studying the genetics of heart valve development in zebrafish. I loved my time at the bench, but there were, of course, many ethical problems that arose in laboratory life. While theological writings were helpful in many ways, nowhere did I find answers to the problems that my colleagues and I discussed: concerns over the pressures of research, anger over the publication restrictions of a Material Transfer Agreement, disputes over authorship, fears of being scooped, or confusions about how to negotiate mentoring relations. Even more troubling, nowhere did I find the question that I and other Christians in the laboratory faced: how can my work serve as a Christian vocation? The literature in theology failed to address for the laboratory what Michael Banner has called the ethics of everyday life.[2]

In part, this lack results from the fact that most of the work on science and religion has dealt with the relationship between scientific theories and Christian doctrines, staying largely at the level of ideas. Yet, as Peter Harrison recently argued, it is a mistake to think of either religion or science as *only* a body of propositional claims, a mistake that ignores the many different historical instantiations of religion and science.[3] Such propositional claims are important, but these endeavors of science and religion are much more than just these propositional claims. We must also consider them as practices.

As Talal Asad argued against Clifford Geertz, we cannot think of religions as merely theoretical constructs of symbolic systems or sets of propositional claims.[4] For most of their adherents, the central aspect of religion is that of practice – not only ritual practice but also all of the small, daily practices that create a whole form of life. Christian ethicists have come to the same conclusions as evidenced by the increasing focus on liturgy.[5] To think of religion *only* in terms of propositional content is to privilege one form of modern Christianity over most other human experiences of religion, a claim that need not deny the important role that propositional systems of knowledge play in religion.

[2] Michael Banner, *The Ethics of Everyday Life* (New York: Oxford University Press, 2014).

[3] Harrison, *The Territories of Science and Religion.*

[4] Talal Asad, *Genealogies of Religion: Discipline and Reasons of Power in Christianity and Islam* (Baltimore, MD: Johns Hopkins University Press, 1993), 27–80.

[5] For an overview of this literature, see Jennifer A. Herdt, "The Virtue of the Liturgy," in *The Blackwell Companion to Christian Ethics*, ed. Stanley Hauerwas and Samuel Wells, 533–46 (Oxford: Wiley-Blackwell, 2011).

Similarly, many scholars studying science have suggested that to only consider science as the system of ideas found in textbooks or even in published journal articles is to misunderstand a large part of what science is. Throughout the last century, scholars, many of whom were practicing scientists such as Ludwik Fleck and Michael Polanyi, investigated science through the lens of practice.[6] They examined what scientists do every day in the laboratory and how these activities relate to the more fully articulated theoretical systems in the published literature. These reflections on the daily practice of science reached their apex in today's science studies literature in the work of Bruno Latour.[7] Thus, we need to consider both science and religion under the lens of practice.

It is especially critical for Christian ethicists to engage science in terms of practice because, as many of these social theorists and philosophers have argued, practices are critically linked to ethics. It is through practices that a certain form of character comes into existence. Practices shape a form of life and, through those daily engagements, shape an individual's dispositions, how one sees and thus reacts to the world. Alternative forms of practice will shape alternative forms of life and thus different forms of character. This book seeks to apply these insights to the practice of science. Different forms of practice can also create friction, so Christian ethics needs to investigate possible conflicts not only between the propositional content of science and religion, but also between the forms of life shaped by the practice of science and the practice of Christianity.

Problems in the Practice of Science

These questions of moral formation require the examination of how one comes to see and to understand the world and how this understanding shapes how one engages the world. They involve the question of the relationship between subjectivity and truth. Conflicts may arise because of the differences in the relationship between truth and subjectivity in alternative ways of knowing. The relationship between truth and subjectivity has become more obvious in science because the current pursuit of scientific truth seems to fail to provide an adequate formation in moral character. Many people in society are concerned about how scientific

[6] Ludwik Fleck, *Genesis and Development of a Scientific Fact* (Chicago: University of Chicago Press, 1979); Michael Polanyi, *Personal Knowledge: Towards a Post-Critical Philosophy* (Chicago: University of Chicago Press, 1962).

[7] Bruno Latour, *Science in Action: How to Follow Scientists and Engineers through Society* (Cambridge, MA: Harvard University Press, 1988).

expertise might serve the interests of power and threaten the natural order, and many religious believers fear that the pursuit of scientific truth may conflict with faithful discipleship. While a secular scientist might be tempted to say so much the worse for theology and religion, this kind of moral formation has serious implications for the practice of science itself.

It is an especially opportune moment to investigate the relationship between the practices of science and religion because of the crisis in contemporary science discussed by many commentators. As Chapter 1 will discuss, this crisis is exemplified by growing problems with fraud, a lack of reproducibility, a slowing of innovation, and the loss of many young investigators. Several writers trace these issues to problems in the institutions and practices of science. While there are many suggested fixes for one or another aspect of this crisis, this book turns from addressing individual aspects of the practice of science one by one to look at what kind of character today's practice of science as a whole is directed toward shaping. Over the last forty years, shifts in policy, law, and cultural norms have encouraged the scientist to become an entrepreneur, to view the knowledge gained through research as a potential resource to patent or commodify in order to start or contribute to a business. This model of thinking of the self as an entrepreneur reflects modes of thinking more widespread in society.

This diagnosis of the problems of contemporary scientific practice points toward a conflict with a Christian form of life. The scientist entrepreneur is able to turn knowledge and organisms into commodities because she already envisions them in a reductionist manner. One of the most important features of modern moral formation that distinguish it from older models is that the relationship between subjectivity and truth is much more strongly mediated by the physical, organizational, and conceptual tools that we use – in other words, our technology. Scientific practice and research tools are embedded in a modern form of a reductionist rationality that dissolves the objects of the world into a mathematical network of forces that can be freely manipulated for human purposes.[8]

[8] Martin Heidegger, "The Question Concerning Technology," in *The Question Concerning Technology and Other Essays* (New York: Harper Torchbooks, 1977), 3–35. See also Max Weber, "Science as a Vocation," in *From Max Weber*, ed. H. H. Gerth and C. Wright Mills (New York: Oxford University Press, 1958), 129–59; Edmund Husserl, *The Crisis of European Sciences and Transcendental Phenomenology* (Evanston, IL: Northwestern University Press, 1970); George Grant, "Knowing and Making," in *The George Grant Reader*, ed. William Christian and Sheila Grant (Toronto: University of Toronto Press, 1998), 407–17; Hans Blumenberg, *The Legitimacy of the Modern Age* (Cambridge, MA: MIT Press, 1985). In *Laudato si'*, Pope Francis makes a similar argument in terms of a technocratic paradigm.

The daily reductionist practice of research in science allows one to see animals and bodies as divorced from greater systems of meaning and purpose and thus open to commodification. These perceptions and this practical attitude contrast with Christian conceptions of Creation that respect the inherent teleologies of other creatures. Given the lack of demonstrable conflict over the propositional content of science and religion, the actual problem between these two domains of approaching the world may lie in the different forms of character to which these practices give rise.

Two Models of Practice

Given these problems in the practice of science, both on its own terms and in relation to Christianity, it is important to understand exactly how practice shapes character and how character can be reshaped through practice. There are two major models of the ethics of practices in contemporary philosophy and social theory, which I will term the Aristotelian and Stoic models, that differ in whether they emphasize social practices or individual practices. The most influential exponent of the Aristotelian model is Alasdair MacIntyre, but one also finds it in other scholars of science such as Polanyi, as well as in Christian ethicists such as Stanley Hauerwas. Discussions of moral formation in religious ethics have focused largely on social practices, arising in part from contemporary appropriations of Aristotelian ethics that tie the development of virtue to the habituation of prerational passions from an early age through the practices of communities dedicated to virtue. This early training leads to the conscious choice of virtuous rather than vicious actions, choices that further develop virtuous dispositions. As I will discuss in Chapter 3, the ideal form of this model is apprenticeship in a craft. This form of Aristotelianism seems to require the individual to dwell in communities that are dedicated to coherent systems of truth and that possess a consistent set of good practices, which many scholars have related to the Church. This account accurately describes scientific training – in its focus on embodied practices, tacit knowledge, and science as a semi-independent community of research – as well as the way external structures can influence character,[9] but, as I will discuss, such an account raises problems for the ethics of science.

[9] Polanyi, *Personal Knowledge*; Thomas Kuhn, *The Structure of Scientific Revolutions*, 3rd edition (Chicago: University of Chicago Press, 1996).

The Stoic model of moral formation, exemplified in the late work of Michel Foucault and in Pierre Hadot, sees practices much more like techniques for shaping subjectivity, reflecting the current use of various practices for self-shaping and the shaping of subordinates.[10] While accepting the role of social practices and structural influences on subjectivity, Stoic ethics looks to how one can fashion one's subjectivity in a different way. Stoicism emphasizes that one must consciously shape one's own initially disordered perceptions and thus one's practical stance toward the world. In so doing, it seeks to integrate seemingly contradictory value systems and social role responsibilities.[11] Along with many other Greco-Roman philosophies, Stoics drew on an array of meditative and ascetic techniques to consciously work on themselves, adjusting their representations and habits in a long process of continuing conversion away from problematic unreflective practices.[12] Through this process, the Stoic sought to become a subject of truth, to translate a coherent system of truths into an *ethos*. Such an ethical system is ideal for working to counteract the negative effects of formation in an exclusively reductionist rationality. While Stoic care of the self can be used by secular philosophies, it also fits well with Christian ethics, since many of the ancient techniques of the self were adapted to Christianity in the form of devotional, meditative, and ascetic techniques that are still central to many Christians' moral and spiritual lives. This book argues that if one lives in a world with deformed social practices that have already formed one in a problematic subjectivity, as in contemporary research, it is necessary to emphasize this Stoic model.

It is important to note that these two models should be understood as ideal types to which I assign certain contemporary thinkers by connecting them to strands of ancient thought on which they draw. Yet it is clear that

[10] This constructive use of Foucault differs from other Christian philosophers and theologians who have productively used Foucault's work to analyze problems in technology and bioethics. These authors use his critique of liberalism but reject his response to the issues he identifies. For example, Jeffrey Bishop says, "Foucault is a master diagnostician, but he is less helpful with therapy." Jeffrey Paul Bishop, *The Anticipatory Corpse: Medicine, Power, and the Care of the Dying* (Notre Dame, IN: University of Notre Dame Press, 2011), 23. For Brian Brock, "Foucault's explicit moral philosophy is of little use in developing a positive theological account of technological development" because he "champions the will." Brian Brock, *Christian Ethics in a Technological Age* (Grand Rapids, MI: Eerdmans, 2010), 145.

[11] Gretchen Reydams-Schils, *The Roman Stoics: Self, Responsibility, and Affection* (Chicago: University of Chicago Press, 2005).

[12] Michel Foucault, *The Hermeneutics of the Subject: Lectures at the Collège de France, 1981–82*, trans. Graham Burchell (New York: Palgrave Macmillan, 2005).

there are Stoic strands in a thinker like Thomas Aquinas, who draws on Aristotle, and there are Aristotelian influences in a scholar like Pierre Hadot. These are labels that allow for convenient theological exploration rather than strict demarcations. The argument here is that both modes of moral formation have important insights, but both need to be modified by each other and, in their purely secular form, by the insights of Christian ethics.

Overview of the Argument

The argument begins by describing the contemporary crisis of science. This crisis is distinguished by at least three aspects. First, there is a growing distrust of the scientific literature. This distrust results not only from an increasing number of retractions due to fraud or from the manipulation of studies by interested parties, but also, and perhaps more importantly, because of the lack of reproducibility of even honestly performed studies. Some of the well-publicized failures to reproduce important studies may be an inevitable part of the scientific process, but many also reflect carelessness and other defects that are common to the practice of contemporary research. A second problem is that there seems to be a lack of innovation even in research that is well done. Despite indicators of falling rates of innovation, however, scientific research is surrounded by hyperbole that shapes policy decisions and public reactions. Finally, there is a growing concern over the future of science since a large percentage of trained researchers do not continue in their fields. The confluence of these issues has led to a flurry of publications, statements, and calls for action from leading members of the scientific community.

The second chapter traces all of these problems to a single source: the growth of entrepreneurial science. This chapter details how, over the last forty years, the structures and practices of sciences have shifted in order to encourage researchers to think of themselves as market actors and their work as the potential seeds for business opportunities. Three character traits of the scientist entrepreneur are clearly connected to the problems constituting the crisis of science. Competitiveness encourages speed and stress, while tempting one with fraud. The need to respond to the short-term incentives of the market discourages long-term innovation and encourages overselling one's results. Finally, viewing knowledge and nature as property intensifies competition and encourages secrecy. The way that all of these problems are tied to the entrepreneurial form of character encouraged in the contemporary scientist reveals the weaknesses of many suggested solutions to contemporary problems. What actually needs to be addressed is the moral formation of the researcher.

Chapters 3 and 4 address the two different conceptions of the way that practices relate to moral formation described above and how they might change character. Chapter 3 discusses character development in science through the work of Michael Polanyi, Thomas Kuhn, and MacIntyre's influential account of moral formation. In this Aristotelian model, one of the fundamental problems in modern moral formation is the loss of a concept of teleology, which leads to moral incoherence. It is this loss that allows one to view the world as material to be manipulated through techniques and to subordinate the goods internal to a practice, like knowledge in science, to an external good, like money. One becomes educated in both teleology and virtue by training in craft practices that combine teleology, education, and authority. This model describes the apprenticeship of scientific training quite well, suggesting that scientific research itself might be a craft that combats the modern fragmentation of character. The chapter then describes a central problem with this model for addressing issues in contemporary science. The commodification of nature required by entrepreneurial science is made possible by a reductionist worldview shaped by the practices of research themselves. The daily practices, paradigms, and tacit knowledge of biological research reinforce the anti-teleological rationality MacIntyre decries, thus shaping one's vision of organisms in ways that allow them to be viewed as discrete, commodifiable entities.

Chapter 4 turns to a different model of practice to remedy the problems in the Aristotelian model. This chapter explores the daily practices that support this form of reductionist subjectivity and the rise of the entrepreneurial ideal. It first describes how embodied practices shape the affects and dispositions of the subject, using examples from my own eight years as a doctoral and postdoctoral genetics researcher. These and other examples suggest that contemporary power works primarily by encouraging individuals to form themselves as certain types of subjects, thus shaping how people use their freedom. Because systems of power encourage certain forms of subjectivity, ethics, defined as attention to one's own character, becomes an important way to counteract problematic models of moral formation.

Chapter 5 turns to the constructive portion of the project by comparing the Aristotelian-inspired and Stoic-inspired responses to the problems of character in science. For MacIntyre, no individual moral reform can occur without reshaping social practice. This solution points toward a withdrawal from society into communities of virtue. Such a move is impossible for biologists because biology forms the researcher in the rationality of a cosmopolitan community of research, and the discipline's advancement requires the resources of the modern state. Foucault, in contrast, sees that the most effective resistance to dangerous deployments of power lies in scientific experts embedded in the apparatus of power. To act as effective agents of resistance, however, scientists must engage in ethics: the development of a conscious relation to the self that he models on Stoic care of the self. By addressing common theological and philosophical criticisms of Stoicism as rule-based, individualist, and neglectful of truth, I argue that Stoicism can be effectively used in a virtue ethics of science. Finally, I describe both secular and religious models of scientific care of the self in the works of Max Weber, Sinclair Lewis, and Antonin Sertillanges. While the secular models aid devotion to scientific truth, Christian care of the self sets one's scholarly work in the context of a wider set of social and religious concerns.

Chapter 6 applies the theoretical tools developed in Chapter 5 to the problems of contemporary entrepreneurial science through the virtue of boldly speaking the truth, *parrhesia*. Speaking scientific truth today is dangerous because corporate and political actors use temptations and threats to neutralize scientific findings that could harm their interests. Philosophical care of the self in antiquity bolstered the virtue of *parrhesia* by transforming the individual into a subject of truth, so this care can encourage the scientist's commitment to truth in the face of risk today. The chapter examines scriptural and historical evidence of *parrhesia* in Christian ethics, showing that Christian faith and practice can deepen this risky truth-speaking by tying it to trust and faith in God. The Christian has broader obligations to both speak moral truth in science and defend the consonance of religious and scientific truth in the face of resistance from both scientific and religious communities.

Chapter 7 explores truth and conversion in more depth by examining the strengths and limitations of practice-based social theories of moral formation for developing a substantive Christian ethics. Many commentators are concerned that investigations of the role of practices in truth undermine the possibility of objective truth, while the malleability of character in the face of practices seems to undermine the possibility of a stable subject. In contrast, this chapter argues that an emphasis on the subjective preparations necessary for an individual to attain truth and the transformations that truth can work on subjectivity not only are compatible with but are essential for Christian ethics. Yet critics are correct that a social theory like Foucault's does not give a substantive ethic because it lacks a system of truth for ethics as well as an ideal form of life that the ethical subject can seek to achieve. These elements are given concrete form in one strand of Christian tradition through a *Logos* theology. The chapter concludes by showing how looking at the ties between truth and subjectivity contributes to a religious anthropology that also emphasizes this tie, the Augustinian understanding of the *imago dei.*

CHAPTER 1

The Crisis in Science

On April 22, 2017, an estimated one million people gathered in cities around the world to take part in a March for Science.[1] People marched to confront grave threats to the current scientific research enterprise: political and economic interests pressuring scientists, widespread denial of well-grounded scientific theories such as evolution and climate change, and general public skepticism of science. In many activist circles, especially since the 2016 election in the United States, most of the attention in this area has focused on these external threats to scientific research. Such external attacks are extremely important to confront, and Chapter 6 will address some of them, but, I will argue, these are not the most dangerous threats to scientific research. An even more insidious danger from within science itself threatens greater long-term damage to the research enterprise. The very practice of research seems broken in many ways. Instead of presenting a clear enemy that can be politically confronted through popular actions like community organizing and protest, these dangers arise from the structures of science, and the way that these structures shape the very character of the contemporary researcher. These problems erode the foundations of the scientific enterprise, endangering its ability to accurately describe the world. A growing list of senior figures in the field, disciplinary societies, national academies, and science journalists are issuing statements, reports, and publications recognizing that there are important issues in how scientific research is currently practiced.[2]

[1] March for Science, "The Science behind the March for Science Crowd Estimates," *Medium* (blog), May 15, 2017, https://medium.com/marchforscience-blog/the-science-behind-the-march-for-science-crowd-estimates-f337adf2d665.

[2] Richard Harris, *Rigor Mortis: How Sloppy Science Creates Worthless Cures, Crushes Hope, and Wastes Billions* (New York: Basic Books, 2017); National Academies of Sciences, Engineering, and Medicine, *Fostering Integrity in Research* (Washington, DC: National Academies Press, 2017); National Academies of Sciences, *Statistical Challenges in Assessing and Fostering the Reproducibility of Scientific Results: Summary of a Workshop* (Washington, DC: National Academies Press, 2016); Bruce Alberts et al., "Rescuing US Biomedical Research from Its Systemic Flaws," *Proceedings of the National Academy of Sciences* 111, no. 16 (2014): 5773–7; Arturo Casadevall and Ferric C. Fang,

In this chapter, I will discuss three of the major internal threats confronting science today.[3]

First, there is a concern among scientists about whether the scientific record is trustworthy. This problem is known as the reproducibility crisis. Most visibly in psychology, but also in a growing number of other disciplines, scholars are realizing that many scientific journal articles are obviously incorrect because other laboratories, and even other members of the laboratories that published the papers, cannot duplicate the reported results. Practically, this failure indicates a massive waste of resources: from money spent on attempts to reproduce the findings, the time and resources exhausted on other research that attempts to build on incorrect findings, to efforts wasted on clinical trials based on incorrect findings. These experiments and research directions are simply useless. Perhaps even more importantly, reproducibility failures lead to a loss of trust in a field and in its investigators. As we will see in later chapters, this kind of trust among investigators for simple correctness in reported findings is essential for research to advance. Lack of reproducibility threatens the very existence of science.

Second, even when results are sound, there are many who question whether they are truly interesting. Every few years, another next-big-thing appears that promises to revolutionize health, like gene therapy, genomics, or regenerative medicine, drawing government funding and journalistic attention. Unfortunately, most of this hype has failed to translate into dramatically superior new cures. In fact, innovation in fields like biomedical science seems to have slowed over the last twenty years. Beyond mere disappointment, these fads in research affect the field as a whole, shaping research funding in ways that deform the overall scientific enterprise and raise patient hopes only to dash them with results that are minimal or slow to appear.

Finally, there are dangers to the sustainability of our research infrastructure. Many people who are trained in research do not end up pursuing careers in academic research.[4] Even more troubling, some of the best

"Reforming Science: Methodological and Cultural Reforms," *Infection and Immunity* 80, no. 3 (2012): 891–6; Ferric C. Fang and Arturo Casadevall, "Reforming Science: Structural Reforms," *Infection and Immunity* 80, no. 3 (March 1, 2012): 897–901.

[3] These, of course, do not exhaust the problems within the practice of science. For example, there are clear issues regarding the lack of representation of women and minorities within scientific communities, varying in extent by field.

[4] National Research Council, *Research Training in the Biomedical, Behavioral, and Clinical Research Sciences* (Washington, DC: National Academies Press, 2011).

junior faculty and advanced postdoctoral researchers in the system drop out of scientific research for other careers. While there are often good reasons for these decisions that do not indicate problems in the field of science,[5] a significant number of these researchers leave because of burnout.[6] A career in research has become incredibly stressful, and the path from postdoctoral fellow to tenured faculty has become very narrow. Research funding has not kept pace with the growth of laboratories, and much of the money dedicated to research flows to senior scholars. If one does not succeed in obtaining tenure, there are few ways for a researcher to continue in the field. These losses are a tremendous waste of the money and years devoted to training these talented individuals. By not providing them more than one type of future in research, the field is squandering the next generation of researchers.

To find solutions to these problems, it is essential that one properly understand them. This chapter will explore each of these three problem areas, which will provide the background necessary to identify their root causes. I will place the greatest focus on deciphering the issues with reproducibility, as this elicits the most interest by commentators and poses the most immediate threat to the scientific enterprise. The other two issues of innovation and retaining the next generation, however, are crucial for the long-term prospects of scientific research.

Reproducibility

The most important characteristic of scientific research is its reliability. Foundational to modern science is that it is objective, meaning that the results of experiments should be approximately the same for any researcher who reproduces the experiments under the same conditions. Ideally, other researchers will replicate published results, thus providing a verification of the existing literature. In practice, such replications are rarely done, for a number of reasons: it is inefficient; sometimes the results are not widely interesting; frequently, the laboratory that wants to draw on the findings is in an adjacent field; and, practically, it is difficult to learn the techniques to exactly replicate the experiment.[7] A commitment to replicating every result that is published would slow scientific research tremendously. As Michael

[5] Many people decide to use their research training in other productive ways, such as in my case moral theology.

[6] Alberts et al., "Rescuing US Biomedical Research from Its Systemic Flaws."

[7] H. M. Collins, *Changing Order: Replication and Induction in Scientific Practice* (London: Sage Publications, 1985).

Polanyi notes, "Nobody knows more than a tiny fragment of science well enough to judge its validity and value at first hand. For the rest he has to rely on views accepted at second hand on the authority of a community of people accredited as scientists."[8] Therefore, the scientific research community functions in large part on trust – trust in other laboratories in your own field and trust in researchers in other fields.

Problems arise when results are not trustworthy. Without complete verification through replication, one cannot know for sure whether an experiment is correct. Peer review was implemented in part to catch poor quality research before its publication, but it can only go so far in the absence of a complete knowledge of the data. Increasingly, it seems that possibly incorrect, unreproducible research survives peer review. This has grave implications for science, since other researchers depend on this literature for the continuation of their projects. This crisis in the reliability of scientific results comes in large part from three factors: fraud, lack of care, and bias.

Fraud and Misconduct

Perhaps the most obvious problem in terms of reproducibility is research fraud. In the most blatant form of scientific misconduct, a researcher might simply make up research results. Anyone who follows scientific misconduct is aware of surprisingly bold deceptions occurring in a wide spread of disciplines. In the heat of the ethical and political debates over embryonic stem cell research and therapeutic cloning, South Korean veterinarian and scientist Hwang Woo Suk claimed to have successfully cloned a human in order to generate embryonic stem cells from the resulting embryo.[9] At the time, this result seemed to validate the promise of therapeutic cloning through a research advance, emboldening those arguing for this research because of its promise of health benefits. Because this seeming research triumph was so far in advance of the rest of the field, Hwang became a South Korean national hero, and the national postal system even made a stamp in his honor. Yet this story swiftly unraveled as investigators found no evidence that the research had been successfully performed and discovered an ethical nightmare of female laboratory

[8] Michael Polanyi, *Personal Knowledge: Towards a Post-Critical Philosophy* (Chicago: University of Chicago Press, 1962), 163.

[9] For overview of this story, see Dennis Normile, Gretchen Vogel, and Constance Holden, "Cloning Researcher Says Work Is Flawed but Claims Results Stand," *Science* 310, no. 5756 (2005): 1886–7.

assistants forced to donate eggs for research. While this fabrication was identified in fairly short order, it had real effects in the world, such as temporarily reinforcing funding for ethically controversial therapeutic cloning research and shaping political and ethical discussions.[10] Hwang was tried and convicted as a result of these exposés, but he has found a home in the biotech industry. His company, Sooam Biotech, specializes in cloning dogs and other animals and has partnered with major Chinese biotech companies on other research ventures.[11]

Almost a decade later, another blockbuster misconduct case occurred in stem cell research.[12] Haruko Obokata and her coauthors claimed to transform normal adult human cells into pluripotent cells, which like embryonic stem cells can become any cell of the adult human body.[13] Starting with Shinya Yamanaka in 2007, many people had already accomplished this feat.[14] What was surprising was the method – merely treating cells with acid or applying physical pressure. Such a simple and cheap method would revolutionize the field, dramatically decreasing the price and difficulty of generating pluripotent stem cells. Obokata became famous, especially in Japan, where she became a spokesperson for women scientists. Yet, again, initial promise turned into public humiliation. Precisely because it was such a simple and surprising method, many other laboratories attempted to repeat the experiment. None was successful, even after they went to the original laboratories and attempted replication side by side with the investigators. The papers were retracted, and a study eventually found that embryonic stem cells must have been mixed with the original tested cells, probably intentionally, but they did not provide

[10] Eventually, a different set of researchers successfully performed therapeutic cloning (Masahito Tachibana et al., "Human Embryonic Stem Cells Derived by Somatic Cell Nuclear Transfer," *Cell* 153, no. 6 [June 6, 2013]: 1228–38), but by the time they succeeded, the field had largely moved on to induced pluripotent stem cells, which are much easier to use, dampening the excitement around the finding. For a discussion of this shift, see Ted Peters, "Stem Cells, Aging, and the Fountain of Youth," *Theology and Science* 10, no. 3 (2012): 221–3.

[11] Drake Baer, "This Korean Lab Has Nearly Perfected Dog Cloning, and That's Just the Start," *Business Insider*, September 12, 2015, www.businessinsider.com/how-woosuk-hwangs-sooam-biotech-mastered-cloning-2015-8; David Cyranoski and Boer Deng, "Stem-Cell Star Lands in Same Venture as Disgraced Cloner," *Nature News*.

[12] For full story, see Dana Goodyear, "The Stem-Cell Scandal," *The New Yorker*, February 22, 2016, www.newyorker.com/magazine/2016/02/29/the-stem-cell-scandal.

[13] Haruko Obokata et al., "Stimulus-Triggered Fate Conversion of Somatic Cells into Pluripotency," *Nature* 505, no. 7485 (January 29, 2014): 641; Haruko Obokata et al., "Bidirectional Developmental Potential in Reprogrammed Cells with Acquired Pluripotency," *Nature* 505, no. 7485 (January 29, 2014): 676.

[14] Kazutoshi Takahashi et al., "Induction of Pluripotent Stem Cells from Adult Human Fibroblasts by Defined Factors," *Cell* 131, no. 5 (November 30, 2007): 861–72.

a full theory. Obokata was found guilty of image manipulation and plagiarism, and her superiors criticized for poor supervision. The outcome was disastrous. The institute where the experiments had been performed, Japan's Center for Developmental Biology, lost much of its funding, and Yoshiki Sasai, the senior author, committed suicide out of shame. Again, this case of misconduct was identified, but not before distracting the whole field for several months.

Misconduct is not limited to research in regenerative medicine. Physicist Jan Hendrik Schon falsified data while working at Bell Labs, leading to at least twenty-eight retracted papers.[15] Dutch psychologist Diederik Stapel invented data for at least fifty-four studies in social psychology.[16] In his case, it was not only other researchers and peer reviewers that were misled. Even his students and coauthors were denied access to the data on which their analyses were based, allowing Stapel a free hand for fabrication. It is a risk in any scientific field that even eminent and trusted researchers can commit fraud.

There is a concern that these major cases of research fraud are only the most obvious examples of a much larger problem. These cases of misconduct were discovered precisely because the implicated studies were so influential or potentially revolutionary and the perpetrators had become important public figures. It stands to reason that fraud in field-specific research by lesser known figures, like the majority of scientific research, is not as likely to be discovered. There are probably many more cases of misconduct because fraud is difficult to detect, frequently involving merely the manipulation of a control image in one figure or the exclusion of results opposed to one's conclusions. As noted, despite the scientific ideal of verification of important experiments by multiple laboratories, most scientific experiments are never reproduced in the classical sense of the term and thus are not tested for their truth value. Even if another researcher does discover discrepancies with the results in a published paper, there is little chance that she would follow up on these issues. A survey of scientists found that while 70 percent had failed to replicate published experiments, only 13 percent had successfully published a report on these failed replications.[17] There is little reward for failing to reproduce another

15 David Goodstein, *On Fact and Fraud: Cautionary Tales from the Front Lines of Science* (Princeton, NJ: Princeton University Press, 2010), 97–102.

16 Yudhijit Bhattacharjee, "Diederik Stapel's Audacious Academic Fraud," *New York Times*, April 26, 2013, sec. Magazine.

17 Monya Baker, "1,500 Scientists Lift the Lid on Reproducibility," *Nature* 533, no. 7604 (May 26, 2016): 452.

laboratory's experiments since journals rarely publish negative results, meaning that many of these problem papers will never be corrected.

This commonsense analysis seems to be backed up by other evidence, such as surveys and reports on retractions. An analysis of survey data found that 2 percent of scientists admitted to committing misconduct themselves, and an additional 14 percent had witnessed misconduct.[18] The number of retracted scientific papers, while still small as a percentage of all papers published, is growing, and the majority of these retractions are due to scientific misconduct of some sort.[19] There has been a ten-fold increase in the percentage of all published articles retracted for fraud since the 1970s.[20] While some challenge the tie between increasing retractions and misconduct, there seems to be a growing problem.[21]

Lack of Care

Yet not all, or not even most, of the reproducibility crisis is due to fraud. In fact, much of the standard, perfectly honest research in many fields of scientific research is also not replicable. Detailed examinations of two fields have revealed these problems. Perhaps most famously and damagingly, in 2011, the Open Science Collaboration led by the Center for Open Science at the University of Virginia initiated a detailed study of the fields of cognitive and social psychology.[22] Assembling a large team of researchers, the project attempted to reproduce the results of 100 studies from three respected psychology journals. In 2015, they published their results: only 36 percent of the results from the replicated studies were statistically significant. Although they used a variety of descriptive measures, in the end, about 60 percent of the experiments were not reproducible. Many of the authors of the original studies challenged the results, citing problems in the replication protocols. Yet this study has led to much concern in the field of psychology over research methodologies. It has also challenged a

[18] Daniele Fanelli, "How Many Scientists Fabricate and Falsify Research? A Systematic Review and Meta-Analysis of Survey Data," *PLOS ONE* 4, no. 5 (May 29, 2009): e5738.

[19] One can find ongoing reporting on scientific retractions, including an attempt to catalogue all retraction notices as they appear, on the blog *Retraction Watch*, www.retractionwatch.com.

[20] Ferric C. Fang, R. Grant Steen, and Arturo Casadevall, "Misconduct Accounts for the Majority of Retracted Scientific Publications," *Proceedings of the National Academy of Sciences* 109, no. 42 (October 1, 2012): 17028.

[21] Daniele Fanelli, "Why Growing Retractions Are (Mostly) a Good Sign," *PLoS Medicine* 10, no. 12 (2013): e1001563.

[22] Open Science Collaboration, "Estimating the Reproducibility of Psychological Science," *Science* 349, no. 6251 (August 28, 2015): aac4716.

number of psychology results that have entered the popular imagination in the last decade, especially ideas of social priming: that what a person sees immediately before making a judgment influences that judgment.

Another field hit by the reproducibility crisis is cancer research. Since Richard Nixon launched the War on Cancer in 1971, cancer has become one of the most heavily funded areas in US biomedical research. Yet, aside from a couple of success stories, such as Gleevec, few therapies have emerged from the molecular dissection of cancer mechanisms. One of the main ways that researchers seek to defeat cancer is by finding drug targets, important proteins in the disease process that pharmaceutical companies can target with therapies. Unfortunately, drug companies found that inhibiting many of these drug targets did not actually inhibit tumor growth in model systems. These failures wasted time and resources. This problem led some in the industry to speculate that the original academic studies were flawed, and some companies began to replicate the original studies before they undertook further research on a target. Two companies made their findings public. In 2011, scientists connected to Bayer published a study showing that they were only able to replicate the results of 20–25 percent of sixty-seven studies they tested, with forty-seven of these studies coming from oncology.[23] Similarly, researchers at Amgen concluded that they were only able to replicate 11 percent of the fifty-three "landmark" studies on cancer that they tested.[24] These failure rates suggest a large problem in the field of cancer biology.

Indeed, John Ioannidis argues that these problems are widespread across science. His statistical analysis suggests that most research findings are false.[25] His findings would mean that $28 billion in expenses on preclinical research alone is wasted every year.[26] In aggregate, these studies indicate a deep problem in the research field.

Finding the proximate causes of these failures is difficult. Commentators have pointed to a number of obvious problems. For example, in cancer biology and biomedical research more generally, there is a problem with

[23] Florian Prinz, Thomas Schlange, and Khusru Asadullah, "Believe It or Not: How Much Can We Rely on Published Data on Potential Drug Targets?," *Nature Reviews Drug Discovery* 10, no. 9 (September 2011): 712.

[24] C. Glenn Begley and Lee M. Ellis, "Raise Standards for Preclinical Cancer Research," *Nature* 428 (March 28, 2012): 531–3.

[25] John P. A. Ioannidis, "Why Most Published Research Findings Are False," *PLOS Medicine* 2, no. 8 (August 30, 2005): e124.

[26] Leonard P. Freedman, Iain M. Cockburn, and Timothy S. Simcoe, "The Economics of Reproducibility in Preclinical Research," *PLOS Biology* 13, no. 6 (June 9, 2015): e1002165.

using the proper cell lines.[27] When performing in vitro cell culture research, it is important to maintain sterile technique to prevent contamination of cells. Most obviously, if bacteria or fungus invade your cells, then they will all be lost. Just as disruptive to good science, but less detectable, is contamination by other cell lines. Certain cell lines grow faster than others, so if you get a stray cell in your dish, then the invader could outcompete all the original cells. Within a few generations, your original cell line will be replaced by the interloper. Since most cells look similar without detailed molecular or histological analysis, you would not even be aware of the replacement. Thus, a breast cancer researcher may think that she is working with a cell line derived from a breast cancer tumor, but instead it may actually be a cell line derived from a completely different type of tumor, or from normal cells, or even from mouse cells. Thus, the perfect breast cancer drug target that she discovers, the targeting of which kills all the cells in her culture, will not actually prove to be a good breast cancer target since the cells were not derived from a breast cancer tumor.

This problem hit cancer biology in the 1970s during the War on Cancer. HeLa cells, to which we will return in Chapter 3, the first immortalized human cell line, grew much faster than other cell lines and, by that time, had contaminated other cell lines used in research.[28] Indeed, many of the cell lines that were isolated were probably just HeLa cells all along. Thus, much of the research emanating from the War on Cancer was useless. While imagining that they were investigating various kinds of tumor types, researchers were in fact just performing research on cells originally isolated from the same cervical cancer. It took a concerted effort to clean up the cell lines in use.

It seems like a similar effort might be required now, since estimates suggest that 20–36 percent of cell lines are contaminated or incorrectly identified.[29] At least today, the task would be simpler, since most cell lines have a molecular tag that researchers can identify through DNA sequencing. Performing such a cell line identification to ensure that one is using the correct cell line before beginning a research project is a simple way to

[27] Harris, *Rigor Mortis*, 93–122.

[28] For discussion, see Michael Gold, *A Conspiracy of Cells* (Albany: State University of New York Press, 1985); Hannah Landecker, *Culturing Life: How Cells Became Technologies* (Cambridge, MA: Harvard University Press, 2007); Rebecca Skloot, *The Immortal Life of Henrietta Lacks* (New York: Crown Publishers, 2010).

[29] "Contaminated Cell Lines Throw Thousands of Studies into Question," Stat, July 21, 2016, www.statnews.com/2016/07/21/studies-wrong-cells/.

prevent confusion and mistakes. Increasingly, authorities and publishers are demanding such authentication. Yet some researchers are resistant to it.

Many of the cases that appear on *Retraction Watch*, a website that tracks retractions and is one of the best sources for ongoing information about misconduct and other research problems, come from similar issues of not testing one's reagents, mixing up data, or losing images. There are also major problems in how statistical analyses are performed.[30] What these examples show is that many of the problems in reproducibility emerge from a lack of care: care in the design of studies, care in the execution of studies, care in the statistical analysis of studies, and care in the repetition of experiments. If one does not take care in experimentation, then one is going to end up with unreproducible results.

Publication Management

Focusing only on individual fraud or individual lack of care misses larger, systeic areas of danger to science. To take a first example, many pharmaceutical corporations have developed an infrastructure to manage the production of scientific publications to support the approval and marketing of new medications.[31] These publication management teams coordinate the research and marketing aspects of for-profit science. Given the potential financial rewards for a new blockbuster medication, the planning of the campaign for approval, physician buy-in, and public excitement must begin with the earliest clinical trials. This managed process starts by organizing clinical trials, which are performed either by academic centers or by Contract Research Organizations (CROs).[32] The trend over time has been to accentuate the involvement of these CROs in

[30] National Academies of Sciences, Engineering, and Medicine, *Statistical Challenges in Assessing and Fostering the Reproducibility of Scientific Results*. Some of these problems in the use of statistics will be addressed in the next section, but I will not provide a detailed examination of these problems because of the technical nature of these discussions.

[31] My description follows the work of Sergio Sismondo, "Ghosts in the Machine: Publication Planning in the Medical Sciences," *Social Studies of Science* 39, no. 2 (April 1, 2009): 171–98. See also Sheldon Krimsky, *Science in the Private Interest: Has the Lure of Profits Corrupted Biomedical Research?* (Lanham, MD: Rowman and Littlefield Publishers, 2003); Marcia Angell, *The Truth about the Drug Companies: How They Deceive Us and What to Do about It* (New York: Random House, 2005); Thomas O. McGarity and Wendy E. Wagner, *Bending Science: How Special Interests Corrupt Public Health Research* (Cambridge, MA: Harvard University Press, 2008); Carl Elliott, *White Coat, Black Hat: Adventures on the Dark Side of Medicine* (Boston: Beacon Press, 2011), 25–50.

[32] For CROs, see Philip Mirowski, *Science-Mart: Privatizing American Science* (Cambridge, MA: Harvard University Press, 2011), 220–39.

the research process. These are companies that are not traditional corporate or pharmaceutical research laboratories because they do not seek to perform any innovative research. They merely hire scientists and doctors, usually on a fairly short-term basis, to perform standardized protocols. The research is sliced into small increments, with different people performing each part of the protocol and others doing the analysis, allowing only the publication management team to have overall control of the interpretation and dissemination of results. Therefore, the research is directly under the control of the company paying for the research. The problem with academic medical centers from the perspective of the pharmaceutical company is that traditional academic researchers want more control over the interpretation of the results and are more likely to challenge a narrative emerging from the company.

After CROs perform the trial, company scientists and statisticians analyze the data. Frequently, one of the many consulting firms specializing in ghost writing scientific articles is then contracted to compose the resulting article. Ghost writers are usually trained as scientists but have then transitioned into the field of technical writing. The company gives the ghost writer the argument of the paper, the figures, and the points to highlight. The ghost writer produces a paper drafted according to these technical specifications. After the paper is written, the company contacts an opinion leader in the field to attach her name to the paper, whose imprimatur would increase the influence of the results. People who might be suspicious of findings from a little-known team of corporate scientists are more likely to trust a paper supposedly written by a thought leader. It is not that the thought leader has no influence on the paper: she may suggest changes to the paper, and these may or may not be taken into account. Still, she will have very little knowledge of the primary data and is ultimately not the true author of the ideas or the paper. After securing the agreement of the thought leader, the publication management team coordinates the submission of the papers and conference presentations with regulatory meetings for drug approval or with marketing campaigns for the release of the drug. Through this apparatus, the company maintains complete control of the messaging around a drug.

In this model of science writing, neither the credit nor the responsibility for a work is any longer tied to specific individuals, which traditionally has been a fundamental function of scientific authorship.[33] These

[33] For discussion of scientific authorship, see Mario Biagioli and Peter Galison, eds., *Scientific Authorship: Credit and Intellectual Property in Science* (New York: Routledge, 2002), 253–357.

developments have forced the International Committee of Medical Journal Editors to ask seriously, "What is an author?" and to develop guidelines for appropriately specifying credit.[34] Even after these guidelines were published in 2005, however, a large industry in publication management remains because the relationship with pharmaceutical companies is extremely profitable for journals as well. Money from publications of large clinical trials, sponsored sections, and reprints purchased for distribution to doctors is critical to journals' financial viability.

Because all stages of the research and analysis are under corporate control, the current model allows for extensive bias in the design and reporting of results. The tools for manipulating results to improve the chances of a successful clinical trial are well known. The company can compare the drugs with placebo controls rather than with the best standard of care, undermining the requirement for true clinical equipoise in clinical trials and making the experimental drug seem better. Companies can manipulate the dosages of drugs by testing lower-than-normal doses of the control, making it appear less effective than the drug under trial. Researchers can use nonrepresentative trial populations to look at the side effects of a drug, testing a drug for the elderly in young people, for example. As Sharon Kaufman notes, a drug whose target population is people in their eighties will lead to far fewer side effects when used by people in their sixties, so by enrolling people in their sixties in a clinical trial, a company increases their chance of success but misrepresents the real-world use of the drug.[35] One can select an endpoint of the trial that reflects a biomarker or is short term rather than an actual long-term health outcome.[36] Finally, a major problem in the field is that statisticians can select and manipulate data to minimize safety concerns and optimize trial results.[37] Without access to the entire history of the experiment and its data sets, some of these manipulations are difficult to detect, but companies have little incentive to make comprehensive statistical information public. Indeed, clinical trial data are frequently classified as trade secrets.

[34] International Committee of Medical Journal Editors, *Uniform Requirements for Scientific Authors* (2005).

[35] Sharon R. Kaufman, *Ordinary Medicine: Extraordinary Treatments, Longer Lives, and Where to Draw the Line* (Durham, NC: Duke University Press Books, 2015), 94–8.

[36] See criticisms by Robert Aronowitz, *Risky Medicine: Our Quest to Cure Fear and Uncertainty* (Chicago: University of Chicago Press, 2015); Joseph Dumit, *Drugs for Life: How Pharmaceutical Companies Define Our Health* (Durham, NC: Duke University Press, 2012).

[37] For detailed discussion of these mechanisms, see McGarity and Wagner, *Bending Science*, 61–79.

The contractual nature of research funding contributes to this control of data. CROs have little interest in independent analysis, and academic researchers who accept corporate funding frequently sign confidentiality clauses that allow companies to block the publication of data that reflect poorly on their product.[38] Such clauses are not only in contracts in which the individual receives funding. Material Transfer Agreements (MTAs), which academic researchers sign when they need a proprietary reagent from a company, also frequently contain confidentiality and publication review clauses.[39] If the researcher does decide to publish unauthorized findings, then she is open to lawsuits for breach of contract, which, even if she successfully defends herself, are expensive and disruptive. Thus, the contemporary funding regime can ensure silence.

A subtler problem is the conflict of interest introduced when a scientist starts a biotechnology company based on his own research. Usually, he is the leading expert in this area and has one of the most active research agendas in the field. Almost inevitably, continuing research in his academic laboratory will overlap with the research of his company. This overlap can give rise to pressures for the academic laboratory research to reflect positively on the aims of the company.

For all of these reasons as well as for the more innocent, subconscious biases that usually infect research, meta-analyses have consistently shown that studies sponsored by an interested party are significantly more likely to show more positive results for a pharmaceutical, process, or device than independent studies.[40] Yet because pharmaceutical companies provide so much of the research funding for clinical trials, panels reviewing these medications, boards making recommendations to the field, and individual doctors continue to rely on this research with little regard to the possible problems.[41] These problems have very serious effects in the world. There are the obvious examples such as the approval of Vioxx by the Food and Drug Administration (FDA), which relied on research in which indications of adverse effects on the heart in clinical trial had been suppressed. Over its many years in use, it resulted in an estimated 80,000–139,000 heart attacks.[42] These examples eventually come to light after much

38 McGarity and Wagner, *Bending Science*, 109–16.

39 Mirowski, *Science-Mart*, 155.

40 Andreas Lundh et al., "Industry Sponsorship and Research Outcome," *The Cochrane Database of Systematic Reviews* 12 (December 12, 2012): MR000033.

41 John P. A. Ioannidis et al., "How to Survive the Medical Misinformation Mess," *European Journal of Clinical Investigation* 47, no. 11 (2017): 795–802.

42 McGarity and Wagner, *Bending Science*, 76.

human suffering, but they make clear the possibility of less obvious cases of ineffective, expensive drugs with unreported side effects affecting human health.

True and False Innovation

In 2015, *Forbes* magazine proclaimed Elizabeth Holmes "the world's youngest self-made woman billionaire."[43] About a decade earlier, when she was nineteen, she dropped out of Stanford to transform the medical industry by introducing a faster, cheaper kind of blood test requiring only a finger stick rather than a needle jab in the vein.[44] The company she founded to produce and administer this technology, Theranos, raised more than $400 million in venture capital and came to be valued at an estimated $9 billion.[45] Here was a classic Silicon Valley and biotech success story: by breaking free of old institutions, a young genius could design a new device that would radically disrupt the medical testing industry. Or so the story briefly ran.

Reality was less exciting. Holmes, it seems, had exploited tremendous hype and her father's extensive business connections to amass financial backing for a technology that never demonstrably worked.[46] Her fall from grace was swift. Forbes changed her company's valuation from $4.5 billion to $0, and in July of 2016, the federal Centers for Medicare and Medicaid Services banned her from operating blood-testing facilities for two years.[47] This story exemplifies many of the problems identified in the last section: biased, secretive research leading to the implementation of problematic technologies, results that were reported but not replicated (although none of these results were ever published in the academic literature), and a lack of care by researchers.

Yet the truly troubling side of this story surfaces when one asks, even if Theranos had delivered on its promises, how significant was its product?

[43] "The World's Billionaires: #435 Elizabeth Holmes," *Forbes Online*, www.forbes.com/profile/elizabeth-holmes/.

[44] Ken Auletta, "Blood, Simpler," *The New Yorker*, December 15, 2014, www.newyorker.com/magazine/2014/12/15/blood-simpler.

[45] Abigail Stevenson, "World's Youngest Female Billionaire – Next Steve Jobs?," September 23, 2015, www.cnbc.com/2015/09/23/worlds-youngest-female-billionaire-next-steve-jobs.html.

[46] The first of these reports was by John Carreyrou, "Hot Startup Theranos Has Struggled with Its Blood-Test Technology," *Wall Street Journal*, October 16, 2015.

[47] John Carreyrou, Michael Siconolfi, and Christopher Weaver, "Theranos Dealt Sharp Blow as Elizabeth Holmes Is Banned from Operating Labs," *Wall Street Journal*, July 8, 2016, sec. Business.

Holmes promised a technology that would transform health care and even the relationship people have to their own bodies. But once you set aside Theranos's self-generated mythology, all that Homes seems to have attempted was to make one area of the already successful and profitable field of medical testing a little bit cheaper and easier for patients. If you do not like having your blood drawn from a vein, Theranos promised, you would never have to feel the sting of a hypodermic needle again. It is not even clear in Theranos's case that a cheaper testing technology would have benefited public health. Contemporary health care, as many have noted, already suffers from excessive testing and subsequent overtreatment.

In attempting to turn humdrum research into a seemingly transformative clinical application by use of hyperbole while lacking a real innovative impact on human health, the case of Theranos exemplifies one of the more problematic aspects of contemporary science. It seems that we are always faced with a new technology or research avenue that will potentially transform human health. When I started as an undergraduate biology researcher in 1998, the field of genetics was on the tail end of the boom in genetic therapy of the 1990s that promised to cure an untold number of human ailments, a boom that ended with the tragic death of Jesse Gelsinger in 1999 during a trial for safety. Yet even though that promise began to fade, excitement ignited over the Human Genome Project, which would give us the "Book of Life," potentially revealing the genes responsible for a host of common human ailments. Supposedly, the results would lead swiftly to the identification of drug targets and gene therapies that would revolutionize health. It all turned out to be much more complicated than that, but governments and venture capitalists invested billions in that promise. As I entered graduate school in genetics in 2001, regenerative medicine's use of stem cells would allegedly rejuvenate worn-out body parts and heal untold numbers of diseases. These promises encouraged voters in states that would soon slash public spending on education, pensions, and other important goods to approve billions of dollars in funding for research, overriding ethical concerns over the destruction and instrumentalization of nascent human life required by embryonic stem cell research. Again, venture capital, private foundations, and universities invested billions of dollars into new regenerative medicine programs and start-ups. Again, the technology has not delivered as much as promised. With the exception of hematopoietic (blood) stem cells, which were already in widespread use before the regenerative medicine boom, twenty years after the isolation of human embryonic stem cells, researchers are closest to proven stem cell therapies for only a few eye conditions.

Most recently, as with so many other areas in our society, researchers have started to repackage earlier trends: the Cancer Moonshot as the new instantiation of the 1970s War on Cancer, CRISPR as an update to genetic therapy, or Precision Medicine as the renewed promise of genomics. It is unlikely that any of these will be any more revolutionary than the earlier developments.

My argument is not that these are worthless avenues of research to pursue. For example, geneticists and evolutionary biologists have gained many insights from genome sequencing, especially from looking at variability between and within species, and CRISPR/Cas9 is revolutionizing basic research. The problem is not even that these developments will fail to improve human health or lead to therapies. The 1990s hype about genetic therapy led to a few treatments, albeit risky ones, for individuals with severe immune disorders, which are very good developments. Research on stem cells, especially adult stem cells like hematopoietic stem cells, are leading to advances. The problem is the hyperbole surrounding these developments; false promises and mistaken hopes are used to override ethical concerns and redirect public funds. It is unlikely that voters and investors would have directed billions of dollars into any of these fields without this kind of publicity. These dreams of wealth and health redirect money from other important lines of inquiry and other important public priorities.

More troubling, this hyperbole comes at a time when actual innovation that improves human health seems to be decreasing by many measures. Government agencies and venture capitalists have now invested tens of billions of dollars of research money in the dream of a bioeconomy, hoping eventually to profit from the cures and new businesses such an economy will produce. Yet while there have been obvious developments in biomedical research over the last forty years, these decades have seen a relative decline in research productivity in terms of therapies, patents, and publications. Only 15 percent of the drugs approved by the FDA during the 1990s were deemed real innovations, while the others were reformulations or copies of existing therapies. Philip Mirowski, an economist and historian of science, has found that the number of these new molecular entities approved each year declined since the mid-1990s, from fifty-three in 1996 to nineteen in 2009.[48] The recognition that society's investments in

[48] Mirowski, *Science-Mart*; Paul Nightingale and Paul Martin, "The Myth of the Biotech Revolution," *Trends in Biotechnology* 22, no. 11 (November 2004): 564–9.

genetics and biotechnology are not producing the predicted therapies has led to a proliferation of new research paradigms: interdisciplinary research, to increase the value of knowledge through cross-disciplinary synergies; translational research, to speed the translation of results from laboratory to clinic; synthetic biology, to create wholly new biological products; and big data, to generate valuable insights from the mass of proprietary biological information. Hyperbole thus seems inversely correlated with actual innovation.

These false hopes are not innocent, as the example of the current for-profit stem cell field shows. Before exploring this example, I must note that there are many excellent scientists doing rigorous and exciting work on stem cells. There are adult stem cell (ASC) treatments that have been rigorously tested. Biotechnology companies are conducting high-quality clinical trials on induced pluripotent stem cells (iPSCs), although these treatments are, as everyone involved admits, years away from the clinic. The dedicated search for true knowledge about stem cells is in process.

Yet there is another side to the stem cell industry, one denounced and feared by those practicing good science. There are a number of clinics that offer stem cell treatments without going through the difficult process of clinical trials. For years, desperate patients have been flying to unregulated offshore clinics to receive embryonic stem cell (ESC) or ASC injections, therapies that have not been tested and that may not even have a valid scientific rationale for treating the disease in question.[49] Even in this country, there are clinics offering to inject patients with stem cells, usually from fat, amniotic stem cells, or the patient's own stem cells, in order to cure an incredible number of diseases, for which there is no clinical evidence, or even real clinical rationale, that stem cells would help. These treatments can run into the tens of thousands of dollars. These for-profit clinics engage in direct-to-consumer advertising and seminars to invite potential patients and then use high-pressure tactics.[50] Reputable stem cell scientists fear these procedures because, beyond their concerns for truth and patients' health, they know that a bad outcome from an unregulated clinic could bring the whole field of stem cells into disrepute.

[49] Douglas Sipp, "The Unregulated Commercialization of Stem Cell Treatments: A Global Perspective," *Frontiers of Medicine* 5, no. 4 (December 2011): 348–55.

[50] Leigh Turner and Paul Knoepfler, "Selling Stem Cells in the USA: Assessing the Direct-to-Consumer Industry," *Cell Stem Cell* 19, no. 2 (August 4, 2016): 154–7.

One of the most dramatic disputes over an untested stem cell treatment was the Stamina Foundation affair in Italy.[51] Stamina claimed to treat a wide variety of illnesses using putative stem cells isolated from bone marrow and then manipulated to become neural stem cells. These treatments were shut down in 2012 by the Italian authorities for safety concerns, but patients' groups desperate for any kind of therapy, even an unconfirmed one, responded with lawsuits and protests in order to receive what they perceived to be their only hope for a cure. Courts started to recognize a right for children to receive the unproven treatment. At this point, Italian stem cell scientists engaged in the struggle against the Stamina Foundation, which they saw as a danger to patients, a struggle that exposed them to protests and threats. The Italian government promised to fund a $3.9 million clinical trial. Stem cell scientists saw the government's investment in a trial as a waste of scarce research funds, another example of the problems of hyped research, but recognized that a trial was better than untested treatments. The study forced Stamina to reveal its methods, which exposed their flaws and ended the plans for a clinical trial. Concerted action by publicly engaged scientists was crucial to ending this potential threat to public safety.

Christian ethicists must be careful when they embrace the seeming promise of new fields of technologies because they can become caught up in the problems of hype over a new field. Through an enthusiasm to be pro-science, theologians can become overly attached to the potential of certain technologies, as the unfortunate embroilment of the Catholic Church with the Stamina saga demonstrates. The Catholic Church has been very supportive of ASC research.[52] Almost every Church document addressing stem cells contains a statement like this one: "Research initiatives involving the use of adult stem cells, since they do not present

[51] For the details of this case, see Elena Cattaneo and Gilberto Corbellini, "Stem Cells: Taking a Stand against Pseudoscience," *Nature* 510, no. 7505 (June 16, 2014): 333–5.

[52] The Catholic Church became involved in these debates because of moral concerns about ESCs. Because the production of ESCs results in the death of embryos, Church statements teach that their derivation violates the negative norm against taking life and is thus an intrinsically evil action. Using already derived ESCs poses the dangers of scandal and cooperation and is thus also condemned. The Church's moral analysis of ASCs is completely different, though, since these cells do not require the destruction of human embryos. Because ESC debates have frequently been portrayed as a battle between religion and science, the Church has used a positive response to ASCs to show that it is not anti-science, that the argument over ESCs is not a choice between science and religion. See United States Conference of Catholic Bishops, "On Embryonic Stem Cell Research" (United States Conference of Catholic Bishops, 2008).

ethical problems, should be encouraged and supported."[53] Note that this is not merely a statement of the licitness of ASC research but an encouragement of it.[54]

The most striking example of this encouragement was the Pontifical Council for Culture's million-dollar contribution to a nonprofit, the Stem for Life Foundation, tied to a biotechnology company specializing in adult stem cell products, Neostem, Inc.[55] This collaboration sought to publicize and encourage research on these cells, because, in the words of the nonprofit's website, they are "proven morally non-controversial sources for regenerative medical treatments."[56] This collaboration resulted in two conferences at the Vatican that were organized by the nonprofit and its parent company and a book cowritten by the president of the company, a priest, and a science journalist.[57]

The Church became tied to the Stamina affair because the Italian senate released the stem cell therapy from regulatory oversight at the same time that the Pontifical Council of Culture held one of its scientific meetings on stem cells in 2013. This conjuncture led the prestigious scientific journal *Nature* to editorialize against the conference, calling it

> a shamelessly choreographed performance. Sick children were paraded for television, sharing the stage with stem-cell companies and scientists desperate to hawk a message that their therapies must be speeded to clinical use ... The scientifically naive Vatican finds the concept of adult stem cells attractive simply because embryos are not involved – yet it ignores the ethical implications of false hope.[58]

Ironically, the Catholic Church's effort to encourage science in the form of ASCs led to a perception of the Church as not only anti-science but also as promoting a false hope in technology.

By false hope, the editors of *Nature* clearly mean that the proposed cures will not work. Of course, Christians must look to a different problem of

[53] Congregation for the Doctrine of the Faith, *Instructions Dignitas Personae on Certain Bioethical Questions* (Nairobi: Pauline Publications Africa, 2009), 32.

[54] By way of contrast, consider CDF statements on genetic technologies. There, too, one form of the technology, germline engineering, is considered illicit, while another, somatic cell engineering, is morally licit. The language used, though, is always that of "in principle morally licit" rather than encouragement (CDF, *Dignitas Personae*, 26).

[55] Mitchell Landsberg, "Vatican Signs Deal to Collaborate on Adult Stem Cell Research," *Los Angeles Times* (October 20, 2011), articles.latimes.com/2011/oct/20/business/la-fi-vatican-stem-cells-20111020.

[56] "Stem for Life," www.stemforlife.org/.

[57] Robin Smith, Tomasz Trafny, and Max Gomez, *The Healing Cell: How the Greatest Revolution in Medical History Is Changing Your Life* (New York: Center Street, 2013).

[58] "Smoke and Mirrors." *Nature* 496, no. 7445 (April 18, 2013): 269.

false hope than the editors of *Nature* – a hope too narrowly lodged in an immanent domain. For years, advocates of various biotechnologies have played on this immanent hope, even in the pages of scientific journals, by promising imminent cures. The statements supporting ASCs emanating from the Vatican or Bishops' Conferences obviously do not encourage patients to engage in medical tourism or unproven treatments. What these examples do show is that the hope engendered by the promise of new research is strong enough to encourage individuals to undergo great risks and to engage in political action to realize them.

Instead of an eager embrace, many new technologies and research areas require caution, which critics have encouraged in response to stem cell treatments. Regulators and legislators have taken action to rein in these for-profit clinics. In 2017, California passed a law providing some consumer protections.[59] The FDA placed these treatments under its jurisdiction and started to issue warning letters to clinics.[60] Yet these attempts at regulation are met with determined opposition. Texas, long a bastion of defenses of a free market attitude toward stem cell treatment,[61] passed more permissive legislation.[62] Other lobbying efforts are attempting to lump ASC treatments with broader attempts to undermine regulatory structures that protect society by ensuring the safety and efficacy of therapies, such as the "Right to Try" movement.[63]

Retention

A third major problem confronting science is the loss of young researchers, which endangers the future of science and suggests a waste of the resources

[59] Michael Hiltzik, "California Moves against Unlicensed Stem Cell Treatments – But Is It Doing Enough?," *Los Angeles Times*, www.latimes.com/business/hiltzik/la-fi-hiltzik-stemcell-regulation-20171013-story.html.

[60] Office of the Commissioner, "Press Announcements – FDA Warns US Stem Cell Clinic of Significant Deviations," www.fda.gov/NewsEvents/Newsroom/PressAnnouncements/ucm573431.htm.

[61] David Cyranoski, "Stem Cells in Texas: Cowboy Culture," *Nature* 494, no. 7436 (February 14, 2013): 166.

[62] Kelly Servick, "Texas Has Sanctioned Unapproved Stem Cell Therapies. Will It Change Anything?," *Science News*, June 15, 2017, www.sciencemag.org/news/2017/06/texas-has-sanctioned-unapproved-stem-cell-therapies-will-it-change-anything.

[63] It is important to note that in the United States, ASC treatments for terminally ill patients are the spearhead of a campaign for "Right to Try" legislation by libertarian think tanks meant to undermine the Food and Drug Administration's power to screen drugs for efficacy. See Paolo Bianco and Douglas Sipp, "Regulation: Sell Help Not Hope," *Nature* 510, no. 7505 (June 19, 2014): 336.

used to fund their many years of training. To take the example of biomedical research alone, only 59 percent of recipients of PhDs in biomedical fields are employed in fields related to biomedical research.[64] There are of course many reasons for this loss of talent.[65] Some simply find promising careers in other fields as their desires and interests for their life change. More tragically, many researchers simply cannot find a job in academia once their training ends. The field has been training more students than it can support for many years now, creating far more qualified candidates than jobs. The field traditionally uses graduate students and postdoctoral fellows as sources of fairly cheap labor in large laboratories. This exchange was debatably fair at a time when these positions were merely transitional stages to an eventual career as a faculty member. Such an arrangement ceases to seem equitable when researchers expect to be postdoctoral fellows for five years, with most individuals jumping from one position to another for even longer. Many never find a permanent position, because the system is not set up to provide permanent employment for those who do not become faculty members.

Perhaps even more troubling are the problems and pressures that emerge in the daily lives of those researchers who actually do succeed in obtaining a faculty position. Contemporary biomedical research requires large amounts of money, and researchers are on their own to secure grants, patents, or corporate partnerships to provide this money for the continuation of their basic research activities. For this reason, the scientific career increasingly becomes a process of turning papers and patents into grants and corporate support. The researcher must sell her vision of future results in return for the necessities of academic survival. Because of this need, grant writing, rather than research or thought about scientific problems, takes up an increasing amount of time for the principal investigator, 30–40 percent of their working time according to some estimates.[66] This requirement leaves little time for reflection on ways forward.

These grants are becoming harder and harder to receive. For example, the National Institutes of Health (NIH) funded only 18 percent of grant applications in 2015, versus a funding rate of 32 percent in 2001.[67] National Science Foundation (NSF) funding success rates also declined,

[64] National Institutes of Health, *Biomedical Research Workforce Working Group Report* (Bethesda, MD: National Institutes of Health, 2012).

[65] See discussion in Alberts et al., "Rescuing US Biomedical Research from Its Systemic Flaws."

[66] Steven Shapin, *The Scientific Life: A Moral History of a Late Modern Vocation* (Chicago: University of Chicago Press, 2008), 245.

[67] National Academies of Sciences, Engineering, and Medicine, *Fostering Integrity in Research*, 101.

but less precipitously. UK success rates are better, typically in the 20 percent range, but still less than ideal.[68] Without receiving one of the premier NIH grants, the R01, it is difficult to obtain a tenured position in biomedical research in US universities. A disproportionate number of the grants that do receive funding go to a few laboratories managed by senior scholars, leaving many junior researchers with little to no funding. The average age at which a researcher first receives an R01 is forty-two.[69] Therefore, even if one does get a first faculty position, one is still under pressure to receive a grant. Reductions in government funding lead to increased competition and reliance on the private sector, with the resulting pressure causing many young investigators to leave science.[70] These pressures continue throughout one's career as grant money is concentrated in a few laboratories. If one fails to find funding at later stages of one's career, then one might have to let go of students or postdoctoral fellows, cut back on research projects, or, in the most drastic cases, close one's laboratory. More importantly, as trainees see the pressure placed on these investigators to obtain this funding, many reevaluate their career goals, questioning whether they would want that life. There are many stages at which people drop out of the scientific community. These aspects of the current structures of scientific funding threaten the future of the research.

Conclusion

Many influential bodies and commentators have recognized that there is a crisis in the scientific community. The National Academies of Sciences, Engineering, and Medicine has released two documents on integrity and reproducibility.[71] A group of important figures, including presidents of bodies and editors of major journals, have released diagnoses of the problem.[72] Science policy advisors are hard at work on the issue.[73]

[68] Dave Matthews, "Decrease in UK Grant Success Rates Prompts Worldwide Comparisons," *Times Higher Education Supplement*, October 6, 2016.

[69] National Academies of Sciences, Engineering, and Medicine, *Fostering Integrity in Research*, 99.

[70] National Research Council, *Research Training in the Biomedical, Behavioral, and Clinical Research Sciences*; Ronald Daniels, "A Generation at Risk: Young Investigators and the Future of the Biomedical Workforce," *Proceedings of the National Academy of Sciences* 112, no. 2 (2015): 313–18.

[71] National Academies of Sciences, Engineering, and Medicine, *Fostering Integrity in Research*; National Academies of Sciences, Engineering, and Medicine, *Statistical Challenges in Assessing and Fostering the Reproducibility of Scientific Results*.

[72] Alberts et al., "Rescuing US Biomedical Research from Its Systemic Flaws"; Casadevall and Fang, "Reforming Science"; Fang and Casadevall, "Reforming Science."

[73] Daniel Sarewitz, "Saving Science," *New Atlantis*, no. 49 (Spring/Summer 2016): 4–40.

Journalists have chronicled these troubles.[74] Many other figures are attempting to find solutions as well.

As concerns that have frequently appeared in this chapter over the problem of competitive science and the role of money in research suggest, these problems are not the result of a few bad apples with poor moral character, a few especially sloppy scientists, or a few overexcited researchers hyping their results to the press. In Chapter 2, I will suggest that the crisis in science arises from sources inside the very structures of science as a result of changes that have impacted the formation of the moral character of the scientist. All of these issues – fraud, reproducibility, hyperbole, lack of innovation, and loss of early stage researchers – can be traced to a shift in the moral ideal of the scientist from a seeker of truth to an entrepreneur that began approximately forty years ago.

[74] Harris, *Rigor Mortis*.

CHAPTER 2

The Scientist Entrepreneur

A recently filed lawsuit may illuminate the contemporary characteristics of the elite scientist, especially as they relate to reproducibility, misconduct, and money. In December 2014, Drs. Piero Anversa and Annarosa Leri sued Harvard Medical School, Brigham and Women's Hospital (and its president), Partners Healthcare System, and Harvard's Dean for Faculty and Research Integrity over an ongoing investigation into possible research misconduct.[1] Dr. Anversa published a number of papers on stem cells in the adult heart, suggesting that adult stem cells might provide a treatment for heart attacks. Despite their impact in the field, other laboratories failed to replicate some of this work, and concerns intensified when *Circulation* retracted a 2012 paper because of corrupt data that Anversa and Leri blamed on a postdoctoral fellow.[2] After receiving complaints about this paper, Harvard began an inquiry leading to a broader investigation of these laboratories.

Anversa and Leri's lawsuit alleged that this investigation was flawed: too slow (the initial inquiry took over a year, as did the full investigation), filled with breaches of confidentiality, and marred by conflicts of interest. Especially fascinating are the details regarding the role of monetary rewards in research that emerged from this legal complaint. The plaintiffs sought damages for the collapse of a multimillion-dollar buyout of their biotechnology company on news of the investigation.[3] Moreover, Mt. Sinai and the University of Miami were pursuing them for "prestigious and lucrative positions" but halted because of the investigation. These researchers were not alone in receiving monetary rewards for research, since their lawsuit

[1] Carolyn Y. Johnson, "Stem Cell Scientist Sues Harvard for Misconduct Investigation," *Boston Globe*, December 17, 2014; *Anversa v. Partners Healthcare System*, No. 1:14-cv-14424-DJC, compl. (D. Mass.R.1.1 December 16, 2014).

[2] Johnson, "Stem Cell Scientist Sues Harvard for Misconduct Investigation," 5; http://circ.ahajournals.org/content/early/2014/04/08/CIR.0000000000000049.

[3] *Anversa v. Partners Healthcare System*, 22.

alleges conflicts of interest throughout the university. A member of the investigating committee as well as the president of Brigham who oversaw the investigation were on the Scientific Advisory Board of a biotech company that competed with Anversa's laboratory and company.[4] These monetary rewards indicated that biomedical science is not just characterized by competition over knowledge but that it actually is a competitive marketplace just like the broader economic market: "Harvard, Brigham, and Partners compete in the marketplace for grant funding, including against Plaintiffs in their capacity as owners of Autologous/Progenital."[5] If this is true, then such problems undermine the reliability of the scientific record as it becomes embroiled in competing corporate strategies, with investigators who use papers to support the development of their own companies faced with reviewers with investments in different companies.

Essentially, this lawsuit shows how the biologist, if he is successful, is now an entrepreneur.[6] This entrepreneurial activity, the involvement in business investments and consultations, is not something that is regarded as controversial or disputed. Rather, it is something that is almost essential if one wants to be at the top of one's field. The scientist must constantly pursue grants in a competitive marketplace. She must seek to turn research results into intellectual property that can secure greater funds through licensing. Ideally, these patents can serve as the basis for starting a biotechnology company, which she can eventually sell to a pharmaceutical corporation. Proceeds from licensing deals or sales of companies can help to keep her academic laboratory going, and research in her academic laboratory can further the ends of the company. If he does not start a company, at least the scientist entrepreneur can consult for or serve on the board of one. In all of these ways, knowledge gained through competition and transformed into property begets material rewards.

This move toward entrepreneurship is not something that is frowned on. Rather, it is positively encouraged by the current structures of science. There is an entrepreneurial ideal in contemporary science. As the example

[4] *Anversa v. Partners Healthcare System*, 13.

[5] *Anversa v. Partners Healthcare System*, 27.

[6] For discussion of the figure of the scientist entrepreneur, see Paul Rabinow, "Severing the Ties: Fragmentation and Dignity in Late Modernity," in *Essays on the Anthropology of Reason* (Princeton, NJ: Princeton University Press, 1996), 129–52; Cathy Waldby and Robert Mitchell, *Tissue Economies: Blood, Organs, and Cell Lines in Late Capitalism* (Durham, NC: Duke University Press, 2006); Melinda Cooper, *Life as Surplus: Biotechnology and Capitalism in the Neoliberal Era* (Seattle: University of Washington Press, 2008); Steven Shapin, *The Scientific Life: A Moral History of a Late Modern Vocation* (Chicago: University of Chicago Press, 2008); Philip Mirowski, *Science-Mart: Privatizing American Science* (Cambridge, MA: Harvard University Press, 2011).

above demonstrates, lawsuits seek to protect these entrepreneurial interests in companies. Major science journals write news features on entrepreneurship.[7] Universities build innovation hubs and found technology transfer offices to encourage faculty to become entrepreneurs. Even so, the government is concerned that the academy is not doing enough to encourage biotech entrepreneurship, so the 2012 US National Bioeconomy Blueprint calls for federal agencies to develop "incentives for academic institutions to enhance entrepreneurship and restructure training programs" so that the country will have a workforce properly formed in entrepreneurial habits for the new bioeconomy.[8]

Moreover, graduate students are exposed every day to professors who have started their own companies, who sit on the boards of large corporations, or who migrate from academia to posts in industry. As many commentators indicate, the most intellectually successful scientists were initially the ones most likely to ally with business and still are.[9] Mentors and those who have achieved the greatest distinction in the field become scientist entrepreneurs, and it thus becomes a desirable path for the field as a whole.

Despite the advantages that entrepreneurship promises, such as increased economic growth and faster movements of discoveries to market, there are also potential dangers tied to this ideal, as the preceding example shows. While it is wrong to romanticize the past (many scientists of previous generations were criticized for their work in the military industrial complex), it is important to note what changes a new moral exemplar and an altered moral training might introduce to research. Indeed, my argument in this chapter is that this new moral ideal in science, the entrepreneurial ideal, is the common root from which the problems described in the previous chapter have grown.

Building on social scientific research on the influence of neoliberal policies on science, this chapter argues that the structural changes that were put in place to encourage scientific entrepreneurship brought about conditions that encouraged the three problems in science outlined in the previous chapter. The increased focus on competition created the need to publish more articles at a faster rate, and thus created a framework

[7] Jeffrey Mervis, "Business Decisions," *Science* 328, no. 6240 (2015): 1190–3.

[8] "National Bioeconomy Blueprint" (The White House, April 2012), 5, www.whitehouse.gov/sites/default/files/microsites/ostp/national_bioeconomy_blueprint_april_2012.pdf.

[9] Sheldon Krimsky, James G. Ennis, and Robert Weissman, "Academic-Corporate Ties in Biotechnology: A Quantitative Study," *Science, Technology, & Human Values* 16, no. 3 (July 1, 1991): 275–87.

incentivizing less careful validation of research findings, a focus on less difficult research projects, and, in the worst cases, fraud. The encouragement to found companies and secure venture capital incited people to hype their research. The pressures brought about by increased competition and other institutional changes disheartened many researchers. The new focus on the entrepreneurial ideal is not the sole cause of these problems of course. As with all social developments, there were many influences. For example, some variations and problems in grant funding that affect these issues preceded and were independent of the neoliberal revolution. The growth of celebrity culture probably shapes some scientists' decisions to hype their results. The glamour of the seeming objectivity of quantification has grown throughout the modern period and has accelerated with the data made available by information technologies, driving judgments of researchers based on factors like number of publications rather than quality of publications. Thus, the entrepreneurial ideal is not the sole cause of these problems, but it is a major influence.

This chapter will describe these shifts by outlining the history of the institutional changes that brought the entrepreneurial ideal into science, especially biomedical research. Since much of this history is not unique to science, this chapter will also explore the broader changes in society that encouraged this entrepreneurial ideal to invade science. After this background, I will examine the problems of this ideal and its associated characteristics as it contrasts with science's classic end of the search for knowledge, connecting them to the aspects of the crisis of science identified in the first chapter. As I will discuss toward the end of the chapter, because the current crisis is rooted in the ideal of the scientist, many of the responses to it suggested by mainstream commentators, such as increased training or surveillance of research, will not be effective. I will draw on this framework in the following chapters to seek a better solution through different understandings of character formation.

The Rise of the Scientist Entrepreneur

As the National Bioeconomy Blueprint's calls for training in entrepreneurship suggest, entrepreneurship was not always central to biological research. As the next chapter will discuss, it actually stands in contrast to prior ideals of the scientist as dedicated to truth rather than to personal enrichment.[10] Many saw the goals of industry and research as conflicting. In the late

[10] Shapin, *The Scientific Life*, 104.

1970s, however, a conjuncture of events brought about this new character of liberal modernity. The scientist was no longer meant to pursue abstract truth for low wages in an ivory tower. Nor was the application of biomedical knowledge to new therapies performed solely for humanitarian reasons. Instead, the new scientist entrepreneur was to move abstract discoveries from his academic laboratory into the marketplace. The scientist's new ideal is to form a biotechnology company based on patents acquired from basic research and money invested by venture capitalists. Rather than shunning material gain for the goods of knowledge, the scientist entrepreneur ideally combines these goods, becoming wealthy from the stock options that flow from his production of knowledge. Science is served by the increased funds made available by harnessing the wealth of venture capital and pharmaceutical companies for basic research; the public is served through the translation of basic knowledge into applied cures; and the nation is served through the creation of new jobs in a new industry that possesses valuable intellectual property.

The rise of the scientific entrepreneurial ideal was not merely an accident of history, brought about unintentionally by other forces. Instead, as Melinda Cooper notes, "The U.S. biotech industry was actively fostered, promoted, and brought into being from the top down by a series of quite deliberate legislative and institutional decisions."[11] Faced with US economic stagnation, increased competition from Europe and Asia, and declining profits in the 1970s, businesses and government increasingly tried to harness research innovations for commercial advantage. It was clear that the United States had some of the most advanced research institutions in the world, so it was thought that through legal and regulatory changes these institutions could shore up the nation's economic advantages. The paradigmatic example of this regulatory change was the US Bayh-Dole Act of 1980, which encouraged universities to establish patents on innovations emerging from federally funded research and to license these patents to corporations for further commercial development. It "effectively gave rise to a new academic personage, the scientist entrepreneur, and a new form of public-private alliance, the joint-venture start-up, in which academics and venture capitalists come together to commercialize the results of public research."[12]

[11] Cooper, *Life as Surplus*, 25. This historical summary will largely trace these changes in science through a case study of the United States because it was the first site of these developments and because the history of these policies is well chronicled. However, many similar changes occurred in other nations, especially the United Kingdom.

[12] Cooper, *Life as Surplus*, 27.

Here there was a move from an ideal of pure science to an ideal of rapidly applying research to economic expansion.[13]

The Bayh-Dole Act was only one of a myriad of regulatory changes that encouraged this new figure. Most importantly, court decisions strengthened intellectual property rights and expanded their scope.[14] In 1980, in *Diamond v. Chakrabarty*, the Supreme Court decided that a genetically modified bacterium could be patented, and, by 1988, Harvard and its researchers received a patent on a transgenic mouse.[15] Through these changes, researchers were encouraged to see their research as possible sources of revenue to exploit. None of this was obvious or inevitable. Until 1975, "Harvard prohibited patents by faculty on medical inventions."[16] Western nations enforced these changes in intellectual property law on other nations through international treaties like TRIPS.[17] This change was brought about through planning and regulation.

Indirect liberalization of financial regulations also enabled scientists to become entrepreneurs by giving scientists the money they needed to start biotechnology companies, beginning in 1976 with Genentech.[18] The opening of NASDAQ, a new stock exchange, allowed scientists to become rich through initial public offerings (IPOs) of their companies' stocks since NASDAQ, unlike other stock exchanges, listed companies that had never shown a profit, like most biotechnology companies.[19] These external institutional changes helped to generate the character of the scientist entrepreneur.

Neoliberalism and the Broader Social Background for Entrepreneurial Science

The figure of the scientist entrepreneur did not arise in a social vacuum but rather was an essential part of broader changes occurring in society,

[13] The ideal of pure science was encouraged by the military, which was the primary funder and manager of the previous regime of science. See Mirowski, *Science-Mart*, 87–138.

[14] Peter Drahos and John Braithwaite, *Information Feudalism: Who Owns the Knowledge Economy?* (New York: New Press, 2003); Christopher May and Susan K. Sell, *Intellectual Property Rights: A Critical History* (Boulder, CO: Lynne Rienner Publishers, 2006).

[15] Mirowski, *Science-Mart*, 139–93.

[16] Mirowski, *Science-Mart*, 117.

[17] Drahos and Braithwaite, *Information Feudalism.*

[18] Shapin, *The Scientific Life*, 215.

[19] Cooper, *Life as Surplus*, 28; Mirowski, *Science-Mart*, 206.

changes in line with the rise of what is generally known as neoliberalism.[20] Support for classical liberalism, which emphasized removing government intervention from the market, collapsed in the aftermath of the Great Depression and the success of fascist and socialist governments throughout Continental Europe. It was replaced by a variety of regimes, such as New Deal liberalism, Christian democracy, and social democracy. All of these regimes enacted a broad system of social welfare supports and engaged in Keynesian policies to manage the instabilities of the economic cycle.[21]

As Michel Foucault describes, these new mechanisms of state control were not meant to squash individual liberty, but to allow it. Over time, however, it became apparent that allowing freedom requires quite a bit of intervention. Due to its mechanisms of power and its rationality, liberalism needs freedom and thus must produce and organize it.

> So, if this liberalism is not so much the imperative of freedom as the management and organization of the conditions in which one can be free, it is clear that at the heart of this liberal practice is an always different and mobile problematic relationship between the production of freedom and that which in the production of freedom risks limiting and destroying it.[22]

The question becomes how the liberal state and society can put in place mechanisms that allow freedom for the disadvantaged without introducing an overly invasive apparatus of bureaucratic intervention. For example, to allow the individual freedom to take part in the market, the income transfers of the welfare state became necessary. To allow the rise of a free market, the government protected domestic production through tariffs on foreign goods. For the market to remain free, anti-monopolistic measures came into effect. Educational institutions created free citizens. The entire armature of the welfare state was necessary to generate spheres of freedom.

[20] The definition of neoliberalism is controversial. For many, it is no more than a catchall reference to contemporary mechanisms of inequality. Others tie it to more specific descriptions of contemporary class warfare (see David Harvey, *A Brief History of Neoliberalism* (Oxford: Oxford University Press, 2007)). Here, I follow an analytic first developed by Foucault, who saw it as a rationality for government based in market mechanisms that encouraged a specific type of entrepreneurial subjectivity in Michel Foucault, *The Birth of Biopolitics: Lectures at the Collège de France, 1978–79*, ed. Michel Senellart (New York: Palgrave Macmillan, 2008). This path has been followed by many others, such as Philip Mirowski, *Never Let a Serious Crisis Go to Waste* (New York: Verso, 2013); Wendy Brown, *Undoing the Demos: Neoliberalism's Stealth Revolution* (New York: Zone Books, 2015).

[21] For a discussion of these historical developments, see Luc Boltanski and Eve Chiapello, *The New Spirit of Capitalism*, trans. Gregory Elliott (New York: Verso, 2005); David Harvey, *A Brief History of Neoliberalism*; Mirowski, *Never Let a Serious Crisis Go to Waste*; Brown, *Undoing the Demos*.

[22] Foucault, *The Birth of Biopolitics*, 63–4.

Yet the managerial apparatus that arose both in the state and in the private economy following World War II drew opposition from both the left and the right. These oppositional elements feared that these mechanisms were too dominating. Throughout the late 1950s and 1960s, cultural critics feared the rise of a drab conformity through strict measures of social control.[23] These concerns over social control culminated in the cultural revolutions of the late 1960s and the early 1970s. From the left, Luc Boltanski and Eve Chiapello describe what they call the "artistic critique" of postwar capitalism, which focused on the need for autonomy and creativity, as allowing for many changes in economic and social organization in favor of entrepreneurialism and flexible labor after the disruptions of 1968 in France.[24] On the right, conservative economists were concerned with government control over the economy and the declines on return on capital. The economic pretensions of the postwar Keynesian consensus were crushed by the economic stagnation and high inflation rates of the 1970s, leading to calls for deregulation. Thus, both the right and left turned against the state in the 1960s and 1970s, causing the rise of what Foucault calls state-phobia.[25] There were at least two parallel forms of this state-phobia: the Frankfurt School on the left, and neoliberalism on the right. In both, "Nazism was at the heart of their reflection," because they saw the postwar growth of the bureaucratic state as echoing the growing power of the German state in the early twentieth century that they argued culminated in Nazism.[26] Both sides easily threw about the word "fascism." Both sides saw the intensification of state power even in Western democracies as leading down the road to serfdom.[27]

Out of these dislocations and debates of the late 1970s, neoliberalism emerged victorious, first through the election of Margaret Thatcher in the United Kingdom in 1979, then the election of Ronald Reagan in the United States in 1980, followed by consolidation of neoliberal opinion in a number of international institutions like the World Bank, International Monetary Fund, and European Union, in the 1980s and 1990s. This school of thought shapes our lives, so it is imperative to understand what

[23] E.g., William H. Whyte, *The Organization Man*, revised edition (Philadelphia: University of Pennsylvania Press, 2002); Sloan Wilson, *The Man in the Gray Flannel Suit*, 4th edition (New York: Da Capo Press, 2002); Herbert Marcuse, *One-Dimensional Man: Studies in the Ideology of Advanced Industrial Society*, 2nd revised edition (London: Routledge, 2002).

[24] Boltanski and Chiapello, *The New Spirit of Capitalism*, 167–202.

[25] Foucault, *The Birth of Biopolitics*, 68, 76.

[26] Foucault, *The Birth of Biopolitics*, 106.

[27] Foucault rejects this analysis, seeing the Nazis and Stalinists resulting from the decrease in state power in favor of nonstate party rule. See Foucault, *The Birth of Biopolitics*, 191.

it is. Neoliberalism was a movement that arose in response to the problems of liberalism after the Great Depression with its foundation in the Freiburg School of anti-Nazi economists, along with Austrian economists such as Friedrich Hayek and Ludwig von Mises.[28] Philip Mirowski specifies this description, characterizing the neoliberal movement as a general theoretical standpoint with many internal disagreements that had its origins in the Mont Pelerin Society, an invitation-only debating club formed by Hayek after World War II.[29] In essence, it is a new conceptualization of liberalism that arose in the crisis of the Depression and in opposition to Keynesianism.

There are three central features of neoliberal thought that are key to understanding the rise of the figure of the scientist entrepreneur. If the criticism of postwar society was that it emphasized broad bureaucratic management that ensured regularity through consensus and considered the subject largely as a consumer of standard mass-produced goods, then the neoliberal vision breaks up bureaucracy through as much competition as possible, encourages the individual to be an entrepreneur, and relies on innovation to disrupt regularity. The importance of these changes is seen best through an understanding of the difference between neoliberalism and classical liberalism. There are two major transformations. First, neoliberals make a decisive move from a model of the market as a site of exchange to a site of competition.[30] Second, unlike in classical liberalism, the market is not a natural object that will occur without intervention, but rather the perfect market is an ideal essence, an eschatological construct that is attained bit by bit through sustained action and adjustment. "It is absolutely not a given of nature."[31] Instead, it is an objective that is produced through state action, which creates laws and institutions that protect competition but that also force more and more areas of life to be understood in terms of market mechanisms. Government bureaucracies, school districts, health care organizations, and charities must be restructured to run on the principles of the competitive market; not in the sense that they must make money, but they must function through competition. Hayek notes that it is not free enterprise at which he aims, but the

[28] Foucault, *The Birth of Biopolitics*, 102–5.

[29] Philip Mirowski, "Defining Neoliberalism," in *The Road from Mont Pelerin: The Making of the Neoliberal Thought Collective*, ed. Philip Mirowski and Dieter Plehwe (Cambridge, MA: Harvard University Press, 2009), 430.

[30] Foucault, *The Birth of Biopolitics*, 118.

[31] Foucault, *The Birth of Biopolitics*, 120.

competitive order.[32] "Neoliberalism should not therefore be identified with laissez-faire, but rather with permanent vigilance, activity, and intervention."[33] The state does not intervene with sure knowledge of how to produce desired social outcomes, as older forms of bureaucratic management did, but only to ensure or to create the continuing conditions necessary for the market.

The market is central because neoliberalism radicalizes the liberal critique of knowledge, which is the idea that no one can plan the ideal set of exchanges because of the sheer complexity of knowledge that would be required. Neoliberals reject technocratic, managerial planning. Yet they also reject almost any possibility of an individual reaching adequate knowledge about any social situation.[34] Instead, they turn to an agent-independent epistemic method, a method for learning the truth that does not depend on the expertise or wisdom of any individual: the market. It is no longer merely an efficient means to process data on preferences to achieve an optimal distribution, but instead the market is thought of as a general information processor, the most effective discovery procedure.[35] In neoliberal hands, the market becomes almost the sole mechanism trusted to produce truth. Many Christian ethicists criticize the move away from justice in the market, but neoliberalism negates the idea of justice in the market in order to support a form of truth: it is not the just price that should be sought, but the true price.[36] No one individual can truly know what is just; one must merely trust the market mechanisms to produce the ideal outcome. In Foucault's terms, the market becomes the site of veridiction, a regime for the production of truth in contemporary thought. One can see this trend in the expansion of the patent regime. By commodifying knowledge, by forcing it into the regime of market competition, neoliberals think that knowledge becomes more efficient, productive, and closer to truth. This trend is also observed in the expansion of futures markets for gambling on the probability of future events, like the now defunct Intrade or the US government's failed attempt to produce a futures market on terrorist actions. Because the market produces truth in the neoliberal

32 Friedrich A. von Hayek, *Individualism and Economic Order* (Chicago: University of Chicago Press, 1948), 111.

33 Foucault, *The Birth of Biopolitics*, 132.

34 Mirowski, "Defining Neoliberalism," 423, 425.

35 Friedrich A. von Hayek, *New Studies in Philosophy, Politics, Economics and the History of Ideas* (Chicago: University of Chicago Press, 1978), 149; Mirowski, "Defining Neoliberalism," 435.

36 Foucault, *The Birth of Biopolitics*, 31.

vision, the government should support the production of markets in as many areas as possible.

For this market ideal to work, there must be subjects willing and able to engage in the market. As many critics have noted, it is not natural for humans to approach every decision as an economically rational market actor.[37] Neoliberals agree – it is essential to neoliberalism to shape people to become such an actor. The proper character type for neoliberal market engagement is the entrepreneur. This focus is new. As Christopher Payne argues, older forms of liberal economics took the entrepreneur for granted, focusing instead on consumption and exchange.[38] Now, though, "The *homo oeconomicus* most sought after is not the man of exchange or man the consumer: he is the man of enterprise and production."[39] Yet because all aspects of life are assessed in market terms, one is not just an entrepreneur at work. Individuals are to engage in competition, investment, and production in all areas of life. For example, the German Ordoliberal policy tried to turn all arenas of life – work, home, school – into small enterprises.[40] Even consumption is reconceptualized as entrepreneurial investment of capital in oneself, allowing for greater acceptance of consumer credit.[41] Thus, the biologist is given the possibility of a monetary stake in her knowledge through a patent, and the institutional framework is changed to make the exploitation of the patent easier. Neoliberalism provides the theoretical justification for the institutional changes in science discussed earlier.

This model of the entrepreneurial subject is seen especially well in the reconceptualization of education. Instead of forming an educated public, education became a form of investment in human capital.[42] This shift has led to dramatic changes in how students and parents relate to the university system. One can choose to invest in one's own capital to increase one's

[37] Peter H. Sedgwick, *The Market Economy and Christian Ethics* (Cambridge: Cambridge University Press, 1999), 35.

[38] Christopher Payne, *The Consumer, Credit and Neoliberalism: Governing the Modern Economy* (New York: Routledge, 2012), 8.

[39] Foucault, *The Birth of Biopolitics*, 147.

[40] Foucault, *The Birth of Biopolitics*, 148.

[41] Payne, *The Consumer, Credit and Neoliberalism.*

[42] Foucault, *The Birth of Biopolitics*, 229; Mirowski, "Defining Neoliberalism," 442; Brown, *Undoing the Demos*; Gary Stanley Becker, *Human Capital*, 3rd edition (Chicago: University of Chicago Press, 1993). The language of human capital has even slipped into encyclicals. "In our time . . . there exists another form of ownership which is becoming no less important than land: the possession of know-how, technology, and skill." John Paul II, "Centesimus Annus," in *Catholic Social Thought: The Documentary Heritage*, updated/expanded edition, ed. David J. O'Brien and Thomas A. Shannon, (Maryknoll, NY: Orbis Books, 2010), para. 32.

future earning potential, but the family can also be conceptualized as an enterprise. Parental time with a child increases the child's human capital, thus increasing the capital of the family enterprise even if it decreases immediate family income. This argument could be applied to eugenics through genetic testing during pregnancy.[43] Since caring for a child with a disability might hinder the productivity of the family enterprise, such children might be selected against even without the external pressures of coercive eugenics programs. Education and genetic selection can be equated through the intermediary of human capital. Therefore, neoliberal biopolitics does not require totalitarian mechanisms to achieve eugenic results. It merely needs to shape conceptions of free actors whose freedom is now deployed in the acquisition of human capital. This new regime changes the problem of eugenics.

In its characterization of the entrepreneur, neoliberalism forces Christian ethics to go beyond its traditional analysis of economic activity. Christian ethics is frequently concerned with the modern subject understood as a consumer focused on achieving private satisfaction. An outstanding example of this analysis is Sedgwick's *The Market Economy and Christian Ethics*, which provides a history of the consumer self and the role consumption plays in contemporary identity.[44] In this analysis, the problem of subjectivity in the market then becomes one of conformity, impulsive desire, and insecurity, with the first two of these reflecting older cultural critiques of the Keynesian economy.[45] The moral response then becomes ethical consumption and simplicity of desire. Such an analysis is of continuing relevance since the consumer does not disappear under this new regime, but it does not address the form of subjectivity most central to neoliberal governmentality, which is the subject of competition: the entrepreneur.

If the entrepreneur is addressed in current literature in Christian ethics, it is in relation to a more casuistic model of how such entrepreneurs should behave in specific circumstances, rather than a theoretical analysis of the meaning of entrepreneurialism for character formation.[46] Such analyses are important, but they do not address the problems created by the entrepreneurial scientist. There are some Christian ethicists who provide a critical

[43] Foucault, *The Birth of Biopolitics*, 227–8.

[44] Sedgwick, *The Market Economy and Christian Ethics*, 82–150.

[45] Sedgwick, *The Market Economy and Christian Ethics*, 272.

[46] For excellent examples of such casuistry, see Albino Barrera, *Economic Compulsion and Christian Ethics* (Cambridge: Cambridge University Press, 2007); Albino Barrera, *Market Complicity and Christian Ethics* (New York: Cambridge University Press, 2014).

analysis of certain aspects of the neoliberal entrepreneurial self. In particular, Christian ethicists are critical of the idea of human capital, especially as it makes its way into understandings of human relations. For example, Sandra Sullivan-Dunbar gives an excellent critique of Gary Becker's model of the family based on its failures to truly capture the nature of familial relations.[47] Yet these discussions of human capital focus on only one aspect of the entrepreneurial self. Therefore, there is a need to give a more holistic analysis of the entrepreneur as entrepreneur.

Finally, these changes do not lead to stable, predictable regularity in society. Instead, the market leads to disruptive technological innovation that ensures perpetual economic growth.[48] Contemporary society does not seek to predict or to manage innovation but rather attempts to produce it by forming entrepreneurial subjects. Neoliberalism works by forming certain kinds of subjects of freedom: entrepreneurial subjects. It requires freedom and produces freedom, of a certain sort. It uses techniques of governance, but they are not the managerial attempts to attain total control. It is a subtler technology that effectively shapes the actions of free subjects.

Entrepreneurial Characteristics and the Crisis in Science

This new moral ideal of entrepreneurship in all arenas of life affects how scientific research is performed, and the effects are not entirely positive. To demonstrate that the contemporary crisis in science is tied to the habits and dispositions of the scientist entrepreneur, I will focus on three of these dispositions that create special problems for the practice of scientific research. First, the entrepreneur must be competitive. Second, the entrepreneur must be responsive to the market. Finally, the entrepreneur must view the world as a collection of things that can be transformed into property that she can sell. In the case of the scientist entrepreneur, this aspect means that both knowledge and fundamental aspects of the natural world must be seen as discrete items open to commodification for sale or investment. These three dispositions put pressure on the integrity of the research enterprise.

[47] Sandra Sullivan-Dunbar, *Human Dependency and Christian Ethics* (New York: Cambridge University Press, 2017), 69–74. See also the discussion of human capital in Sedgwick, *The Market Economy and Christian Ethics*, 163–5.

[48] Foucault, *The Birth of Biopolitics*, 231.

The Competitive Scientist

As discussed in the previous section, one of the hallmarks of neoliberalism is its heavy focus on competition, which transforms markets from places of exchange to places of competition. This competition helps markets to function efficiently by driving down costs, ensuring that market actors exploit all opportunities and knowledge available, and serving to enforce rational discipline. As neoliberal ideas entered research, science became reconfigured as a place of competition. It would be naïve to claim that research was filled with friendly cooperation in prior regimes of scientific organization. It is not that all the elements of this neoliberal assemblage were new practices in and of themselves. It was the conceptual framework in which these elements were situated that was new. Biologists had made money from patents before, but it was never before placed in such an entrepreneurial framework. Biologists had certainly competed before, as one can easily learn from reading about the race to discover the structure of DNA or to decipher of the genetic code, but the stakes became different.[49] It was not money, but prestige, making the effects different.

Competition takes place in science on a number of fronts. In any hot field of research, even if it is not directly marketable, many laboratories work on very similar problems. Whichever laboratory gets the results first gets a major publication. The other laboratories, scooped by the first laboratory, may end up with less prestigious publications or none at all. The other laboratories may consider their investment of time, effort, and resources wasted, even though the simultaneous confirmation of results is valuable to science as a whole. Publication is a winner-takes-all matter.[50] The loss of these publications matter in other aspects of the scientific career because researchers use the number and prestige of their papers to compete on other grounds. As described below, many universities rely on bibliometrics, quantification based on the number of publications a person has and the impact factor of the journals in which these publications appear, to determine promotions, tenure, and raises. There are many problems with such systems of evaluation, but as they stand, they are a great incentive to compete to produce papers. Number and impact factor of papers are also important in grant reviews as part of the judgment of the qualifications of

[49] James D, Watson, *The Double Helix: A Personal Account of the Discovery of the Structure of DNA* (New York: Atheneum, 1968); Lily Kay, *Who Wrote the Book of Life?: A History of the Genetic Code* (Stanford, CA: Stanford University Press, 2000).

[50] Arturo Casadevall and Ferric C. Fang, "Winner Takes All," *Scientific American* 307, no. 2 (2012): 13.

the candidate. Researchers need hundreds of thousands of dollars to keep their laboratories running, so they are continually applying for grants. Yet as the last chapter demonstrated, the competition for grants is intense since the NIH currently funds only 18 percent of applications. If a researcher does not receive a grant, then his institution may deny him tenure or force him to close his laboratory. Publications lead to grants, but grantsmanship itself is a fiercely competitive craft.

The first problem is that such competitive pressures create an incentive structure that rewards fraud. As Diederik Stapel, the prolific fabricator mentioned in Chapter 1, says, "There are scarce resources, you need grants, you need money, there is competition ... Normal people go to the edge to get that money. Science is of course about discovery ... But it is also communication, persuasion, marketing. I am a salesman."[51] Rather than an aberration, he views himself merely as an accentuated version of the contemporary entrepreneurial researcher. One might be tempted to dismiss such claims as the self-justifications of a man who engaged in extensive public deception, except that so many other commentators also point to competitive pressures as central to the crisis of reproducibility and integrity in science.[52] When one is under intense competitive pressure, it is tempting to engage in research misconduct, especially in less blatant forms: retouching a picture of a gel so the results are more convincing or reusing a better picture from another experiment. Pressures encourage fraud. More than that, competitiveness as a personal trait inclines one toward fraud in order to win.

Competitiveness also encourages the sloppy research practice that is at the heart of the reproducibility crisis. When a laboratory needs a paper for a grant, it is tempting to toss out some outliers so that the results are statistically significant. When a researcher desperately needs another paper for the tenure file, it is tempting to go through old data to find a result, any result, despite bad statistical practice. If you fear that another laboratory may soon publish on a possible cancer target, then you will publish the paper without reproducing and confirming the results in a variety of different systems. As researchers rush, they make mistakes and overlook

[51] Yudhijit Bhattacharjee, "Diederik Stapel's Audacious Academic Fraud," *New York Times*, April 26, 2013, sec. Magazine.

[52] Bruce Alberts et al., "Rescuing US Biomedical Research from Its Systemic Flaws," *Proceedings of the National Academy of Sciences* 111, no. 16 (2014): 5773–7; Ferric C. Fang and Arturo Casadevall, "Competitive Science: Is Competition Ruining Science?," *Infection and Immunity* 83, no. 4 (2015): 1229–33; National Academies of Sciences, Engineering, and Medicine, *Fostering Integrity in Research* (Washington, DC: National Academies Press, 2017).

problems. The end at which competitiveness aims is victory, not truth. Thus, the competitive nature of the scientist entrepreneur is intimately connected to the reproducibility crisis in science.

While neoliberalism claims that competitiveness should encourage innovation by driving people to work harder, come up with new ideas, and develop untapped research niches, it can also slow down the progress of science because it encourages secrecy. If a researcher is in a race with other laboratories or has new suggestive results, she is less likely to share them with other researchers because she would be sacrificing a competitive advantage. From a strategic standpoint, it is much better to sit on the results until her laboratory can work through them rather than to discuss them with others who might scoop her. Sharing results too early would be silly from an entrepreneurial standpoint.

Such a response is strange from the standpoint of older models of science. One of the classic Mertonian norms of science is communism or communalism, the shared access to research results by all scientists.[53] This norm is important because shared knowledge is more productive. Each researcher can take advantage of the new information and insight in her own project. The importance of sharing has been intensified by the growing complexity of research projects. Increasingly, innovative research requires collaboration with people from adjacent disciplines and those in one's own field with complementary expertise. Yet such a norm for sharing data makes less sense now that knowledge is a resource to own. In a market understood in terms of knowledge, information is central to competitive advantage and profit. As in the stock market, if everyone knows the same information, then there is no chance to profit. Research results are proprietary information that must remain confidential, obstructing the free flow of knowledge on which science depends. As competition for grants and venture capital increases, it makes less sense to collaborate, even though "the importance of teamwork in science has never been greater."[54] In this way, the secrecy encouraged by competition impedes scientific innovation.

Finally, engaging in a constant competitive struggle wears on researchers. Most scientists entered research in order to pursue knowledge and to help others through discoveries, rather than to engage in competition. They must force themselves to become competitive. Frequently, many of my undergraduate classmates who aimed at a research career bemoaned

[53] Robert K. Merton, *The Sociology of Science* (Chicago: University of Chicago Press, 1973), 273–5.

[54] Casadevall and Fang, "Winner Takes All."

their early biology classes because they were structured around the competition for medical school slots. The pressures created by this competition for grants, patents, and papers, and more so the need to become competitive oneself, are a major aspect of the problem of retention for young scientists.

Market Responsiveness

Another disposition required of the entrepreneur is responsiveness to the demands of the market. One achieves entrepreneurial success by adapting to the open niches in the market, ditching failing aspects of one's enterprise, and abandoning unpromising lines of development. The anthropologist Caitlin Zaloom gives an especially compelling description of this aspect of entrepreneurial character in her ethnography of commodities traders.[55] They must constantly work to erase any attachments they might have to particular trades. Instead, they follow the market, abandoning positions when they start to decline or selling when they turn an initial profit. As she portrays it, they undergo a process of detachment from the things of this world so that they might give themselves over wholly to the movements of the market. These are short-term traders, so they occupy a special relationship to the market, but scientists, too, must be responsive to the market. They must develop research trajectories that can attract venture capital if they seek to start a biotech company. They must follow the hot areas of research that are well funded if they want to receive grants from companies or government funding agencies. These incentives are meant to drive scientists to be more productive of the public good in terms of public demands and economic growth.

Yet a commitment to responding to the demands of the market, shaping oneself to see the world from the perspective of market developments, can also create problems for scientific research. These problems occur most obviously when one seeks not just to perform certain kinds of research desired by the market but to produce certain results desired by market actors. The drive to produce a finding that a funder prefers is one of the reasons that research funded by a company is significantly more likely to produce results that reflect positively on that company's product.[56]

[55] Caitlin Zaloom, *Out of the Pits: Traders and Technology from Chicago to London* (Chicago: University of Chicago Press, 2010), 127–40.

[56] Andreas Lundh et al., "Industry Sponsorship and Research Outcome," *The Cochrane Database of Systematic Reviews* 12 (December 12, 2012): MR000033.

Such bias can emerge from misconduct or carelessness, but just the knowledge of what the funder's preferred result would be can influence one's analysis of data. Bias occurs in a major way through the publication management strategy, where one serves as a single cog in a research and marketing team to produce the one aspect of the project desired by the funder, without engaging with the accuracy of the project as a whole. Thus, market responsiveness can lead to problems in terms of bias and conflict of interest.

This market responsiveness can lead to less obvious problems in terms of innovation because it alters the kind of research performed. If the goal of entrepreneurial research is to generate knowledge to patent, to found a company, and then to sell to others, then the entrepreneur may downplay slower yet more innovative basic research that will not turn into anything marketable for decades in favor of research that he can bring to market in the near future. Much has been written about the increasing turn to short-term profits over long-term investment throughout the economy. People prefer taking an immediate profit because of the uncertainty of the future in the rapidly changing world of the neoliberal economy. In science, this might mean that one forgoes important long-term research projects that would extend beyond the current grant cycle or performance review because they would entail too much risk.

Further, the emphasis might turn to tools and procedures, things that are easy to patent and that companies can sell to researchers, doctors, and others. A feature in the journal *Science* on scientists' entrepreneurial experiences discusses how Jackie Ying, previously at MIT and now director of an institute in Singapore, encourages students and faculty to pursue commercial research and to better envision themselves as responding to the market. As director, she "coax[ed]" one faculty member to switch from research on epigenetics, a fascinating field that explores the influence of the environment on inheritance but that currently has little commercial potential, to tissue engineering, a much more directly marketable field. Even once she switched to a more commercially viable field, the faculty member still pursued an overly ambitious agenda. "Ying also weighed in when the researcher began to think about developing an artificial kidney. 'That would take a very long time,' she told her ... So Ying persuaded her colleague to instead turn ... cells into a toxicity assay for potential cancer drugs."[57] Such cells can be sold to pharmaceutical or cosmetic companies.

[57] Mervis, "Business Decisions," 1191–2. Note that the criticism was based not on the scientific viability of the project but on its commercial viability.

The entrepreneurial imperative forced the researcher from ambitious long-term projects to a short-term minor research advance useful for industry.

Over the last decade, the NIH has also pursued short-term goals through its funding of translational research.[58] This focus on translation arises in response to concerns over the rather meager output of cures and treatments that have resulted from the billions of dollars invested in biomedical research in the last thirty years. By specifically funding translation, the NIH hopes to overcome barriers between the laboratory and the clinic (and the boardroom), but these programs also change the kinds of research pursued. They encourage research that leads to immediate translation into therapies and marketable products.

Many companies and foundations are taking steps to mitigate the problems resulting from cautious, short-term funding. Following the example of German industrial cartels and Bell Labs, Google uses some of the profits from its near-monopoly on Internet advertising to fund long-term research at its Google X center and its biotech start-ups Verily and Calico. Yet even Google expects that these research applications will make money within a few years, offering employees bonuses to quickly kill projects whose payoffs might be too distant.[59] In contrast, the Howard Hughes Medical Institute (HHMI), in Chevy Chase, Maryland, has pursued a different strategy from that of the government or Google, by funding people instead of projects. In doing so, the HHMI has made it possible for scientists to follow their interests rather than cleave to detailed grant plans.

Yet they, too, are turning to a type of market- and technology-based research strategy. The HHMI has started Janelia Research Campus, a facility outside Washington, DC, in Ashburn, Virginia, that is promisingly meant "to recreate the close-knit feeling of legendary laboratories such as the Laboratory of Molecular Biology in Cambridge, UK, where well-funded investigators free of grant-seeking pressures work in small groups."[60] But even Janelia puts its primary focus on new technologies, not basic research, and even when scientists are freed from the restraints of a short-term grant, they are still imprisoned by the assumption that new technology is the goal.

58 S. H. Woolf, "The Meaning of Translational Research and Why It Matters," *JAMA* 299, no. 2 (January 9, 2008): 211–13.

59 Conor Dougherty, "They Promised Us Jet Packs. They Promised the Bosses Profit," *New York Times*, July 23, 2016, www.nytimes.com/2016/07/24/technology/they-promised-us-jet-packs-they-promised-the-bosses-profit.html.

60 Jocelyn Kaiser, "Janelia Farm to Recruit First Class," *Science* 306, no. 5694 (2004): 210–11.

Because of this focus on short-term marketability and immediate technological implementation, many scholars warn that this focus on entrepreneurship in research could endanger the very innovation on which entrepreneurship relies. As a British government report on innovation that otherwise praises entrepreneurship puts it, "System-wide rollouts of ideas like the entrepreneurial university may pose substantial risks . . . The new incentives for commercialization risk crowding out the basic research that expands the boundaries of *future* innovation and economic growth, with unknown long-term effect."[61] It is exactly because the older ideal of the scientist did not aim at profit that he produced the seeds of future innovation.

Finally, this need to envision the world from the perspective of the market feeds another problem connected to innovation: the false innovation promised by hype. If there is not a market for the product one has, then a good entrepreneur will create a market. Through their statements, presentations, and press interviews, the best scientist entrepreneurs, the senior figures, are able to create markets of public funding and venture capital by encouraging inflated hopes about certain lines of research. As Stapel said earlier, "Science is of course about discovery . . . But it is also communication, persuasion, marketing. I am a salesman." Thus the human genome project is not merely an interesting and important research project, but it is marketed as the "Book of Life." Francis Collins argues that precision medicine will interpret this "language of life," thus revolutionizing medicine.[62] These are all ways to attract funding and create a new market. Yet these marketing tactics distort research, focusing money in certain areas due to inflated promises.

Knowledge and Nature as Property

Scientific entrepreneurship also encourages one to see knowledge and nature as property. To be a successful entrepreneur, one needs something to sell, one needs capital. In science, it is knowledge, especially the knowledge of nature that provides those resources. For this reason, the reconstruction of the intellectual property landscape was absolutely crucial for entrepreneurial science to emerge. At first glance, it is odd to think of

[61] Annual Report of the Government Chief Scientific Adviser, "Innovation: Managing Risk, Not Avoiding It. Evidence and Case Studies," 2014, 20, www.gov.uk/government/uploads/system/uploads/attachment_data/file/381906/14-1190b-innovation-managing-risk-evidence.pdf.

[62] Francis S. Collins, *The Language of Life: DNA and the Revolution in Personalized Medicine* (New York: Harper Perennial, 2011).

scientific knowledge as property. How is knowledge of a gene sequence property belonging to a person? Indeed, it seems as if courts are now revisiting the intellectual property landscape.[63] Still, the current intellectual property regime has changed how science is done.

Intellectual property intensifies the competitive nature of contemporary science. Patents emerging from research are one of the primary means that researchers and universities can accrue wealth in the new entrepreneurial setup. If a researcher strikes it big, then she can make an incredible amount of money from a discovery. Stanley Cohen and Herbert Boyer brought in $255 million in licensing fees to themselves, Stanford, and UCSF for the patents to the basic tools of recombinant DNA technology.[64] Researchers need to have good intellectual property to start a biotechnology company. If a discovery is truly novel, then there is generally not much trouble with competition. However, frequently intellectual property does not spring from a truly unexpected or original discovery. Instead, research in the field suggests that there is a result or finding out there that might be valuable, leading many laboratories to try to be the first to discover it. In these cases, a competitive race ensues. Many stories are told in laboratories of the devious methods used in the early days of genetics to be the first to isolate, sequence, and patent a gene, such as people writing down DNA sequences presented at meetings and phoning the information back to their own laboratories in order to exploit it as soon as possible. Whoever is first past the post gets priority, so, as noted earlier, a winner-takes-all system encourages competition.

If there are disputes over when intellectual property arose or who contributed to it, they can lead to massive lawsuits. Sometimes these occur between individual scientists and their institution, such as the famous lawsuit between Genentech and UCSF over human growth hormone. UCSF claims that a researcher took DNA that he had isolated while employed by the university to the company when he left, and that the university should have a stake in the patent and the profits. UCSF was eventually awarded $200 million.[65] These are jurisdictional disputes.

[63] For example, in the 2013 decision in *Association for Molecular Pathology v. Myriad Genetics*, the US Supreme Court struck down the broad patent granted for the genetic sequence to the BRCA 1 and 2 genes, which predispose a person to breast cancer.

[64] Sarah Zhang, "How the CRISPR Patent Dispute Became So Heated," *The Atlantic*, December 6, 2016, www.theatlantic.com/science/archive/2016/12/crispr-patent-in-court/509579/.

[65] Rex Dalton and Quirin Schiermeier, "Genentech Pays $200m over Growth Hormone 'Theft,'" *Nature* 402 (1999): 335.

Frequently and more consequentially for the practice of science, disputes occur between different groups competing for the same results. The most famous recent example involves the CRISPR/Cas9 gene-editing technology. CRISPR, which is revolutionizing genetic research and may allow easier genetic interventions in humans, is already embroiled in a debate over who owns the technique because multiple researchers claim priority to its discovery. A group led by Jennifer Doudna and Emmanuelle Charpentier originally determined how to transform a bacterial antiviral system into a technology, and a group led by Feng Zhang optimized it and used it in mammalian cells. While the Patent Office granted Zhang a patent on the technology, it appears that Doudna and Charpentier had filed a patent application earlier. The most recent rulings have favored Zhang's patent, but the legal cases continue.[66] The resolution of this conflict is quite important because "the CRISPR patent applications filed thus far are drafted quite broadly . . . The patents are likely to prevent even the most basic uses of the CRISPR-Cas9 system without a license."[67] Depending on how licensing agreements are structured, they may or may not impede the advance of research.

The importance of intellectual property also encourages secrecy in a way that harms reproducibility. Access to the full data set used in the experiment as well as all the analytical tools is essential to reproducing results.[68] Yet researchers are frequently reluctant to give full access to data because they hope to continue to mine that data in the future. Giving away data is like giving away one's future research. Such concerns are especially salient for pharmaceutical companies, which have generally tried to retain clinical trial data as trade secrets, release of which would harm their competitive profile in the marketplace. Lack of access to data makes it difficult to check whether the analyses hold up. Further, many of the tools used to analyze the data are also proprietary, either held by companies as intellectual property or developed by researchers as an essential part of their research program. For example, many researchers are loathe to release the source code of the analytic tools they use to mine genomic data. Without access to the full code, though, it is impossible for outside researchers to determine if the analytic tool works properly and ensure the accuracy of the results.[69]

[66] Heidi Ledford, "Bitter CRISPR Patent War Intensifies," *Nature News*, 10.1038/nature.2017.22892.

[67] Zhang, "How the CRISPR Patent Dispute Became So Heated."

[68] National Academies of Sciences, Engineering, and Medicine, *Fostering Integrity in Research*, 7–8.

[69] Roger Peng, "The Reproducibility Crisis in Science: A Statistical Counterattack," *Significance* 12, no. 3 (June 1, 2015): 30–2.

Journal editors, funders, and societies are working to correct these problems by requiring the release of all data, as will be described later, but their efforts push against the incentives of the contemporary competitive environment's demands for secrecy.

Here at the discussion of intellectual property, we also begin to see some of the broader negative impacts of the entrepreneurial ideal. Celia Deane-Drummond has addressed these issues specifically related to genetic patents at length in her *Genetics and Christian Ethics*.[70] While she addresses more conceptual issues involving concerns about playing God, her primary focus is on the ways that patents violate the interests of social justice.[71] Such patents threaten to take a general public good and to place it in private hands, resulting in a system "unfairly skewed towards benefiting a narrow section of the human community."[72] It benefits entrepreneurs and citizens of wealthy nations at the expense of the poor of the world, from whose nations' genetic patrimonies many of the valuable gene sequences come.

There are also broader concerns about this view of knowledge and biomedical tools as property. Animals and the human body become resources from which to extract profit. Many scholars have noted the increasing commodification of research animals, human tissue, research subjects, women's eggs, etc.[73] Researchers lose sight of the human and biological realities of these bodies, and their other purposes and meanings, in the push for commercialization, just as other natural resources fall prey to extractive industries. As will be discussed in Chapter 3, this way of seeing the world is not new, nor does it arise merely from an entrepreneurial subjectivity, but the entrepreneurial framework intensifies it.

Caveats

This discussion suggests entrepreneurial science can contribute to great problems for the field of scientific research. It is important to note what is not claimed here. First, nothing here states that business is antithetical to

[70] Celia Deane-Drummond, *Genetics and Christian Ethics* (New York: Cambridge University Press, 2005), 160–90.

[71] Deane-Drummond, *Genetics and Christian Ethics*, 178–90.

[72] Deane-Drummond, *Genetics and Christian Ethics*, 179.

[73] Rabinow, "Severing the Ties: Fragmentation and Dignity in Late Modernity"; Waldby and Mitchell, *Tissue Economies*; Cooper, *Life as Surplus*; Jill Fisher, *Medical Research for Hire: The Political Economy of Pharmaceutical Clinical Trials* (New Brunswick, NJ: Rutgers University Press, 2008); Mirowski, *Science-Mart*; Melinda Cooper and Catherine Waldby, *Clinical Labor: Tissue Donors and Research Subjects in the Global Bioeconomy* (Durham, NC: Duke University Press, 2014).

scientific research. Businesses have been closely involved in scientific research for the last 150 years. Indeed, some of the most innovative scientific institutions were tied to businesses. One thinks of the Kaiser Wilhelm Institutes in pre-Nazi Germany, which received much of their funding through close links to German business cartels.[74] Or one could point to Bell Labs, the research department of AT&T, which was one of the greatest research and engineering endeavors in world history.[75] What is criticized is a specific form and structure of research that contributes to a certain character of the researchers. These older laboratories left much more freedom for exploration, had much longer time horizons for research, and performed evaluations in very different ways.[76] Just as the general style of corporate management has changed in the last forty years, so has the management of corporate research. The monopolistic, highly regulated AT&T of the 1940s is very different from the similarly situated Google, and so are their research structures. Formerly, pharmaceutical companies did most of their research internally, but now they rely on purchasing smaller biotech companies started from universities with interesting new drugs based on academic intellectual property and then validating these by contracting with CROs. It is the organization of the scientific process that shapes the scientist that is important rather than merely the involvement of monetary interests. It is clear that business will always have an important role to play in translating scientific research into public benefits. The present argument is that the current conceptualization of how that role is performed blocks those public benefits from appearing.

The second caution to note is that this argument is not an absolute rejection of the entrepreneurial subjectivity as envisioned by neoliberalism. It may be that it is an accurate and valuable understanding for how people should operate in the sphere of business. One could imagine critiques of this model, but it is not out of the question that, properly modified, it would be a valuable way to understand a character ethics of the Christian businessperson. The criticism leveled here and elsewhere is that the entrepreneurial subject has become a blueprint imposed on an increasing number of practices and spheres of life. As more spheres of life become understood in an economic model, the entrepreneur becomes the model subject in each of them. People call for entrepreneurial medicine or

[74] See discussion in Mary Jo Nye, *Michael Polanyi and His Generation: Origins of the Social Construction of Science* (Chicago: University of Chicago Press, 2011), 52–66.

[75] Jon Gertner, *The Idea Factory: Bell Labs and the Great Age of American Innovation* (New York: Penguin Books, 2013).

[76] Shapin, *The Scientific Life*, 127–64.

entrepreneurial humanitarianism. This is a mistake. As many have noted, an entrepreneurial model works poorly in settings like international development.[77] It also works poorly in science.

The Limitations of Mainstream Responses to the Crisis of Science

The framework discussed here that explains the crisis of science through the character traits of the moral ideal of the contemporary scientist and the way that the research enterprise is structured suggests why many proposed responses will not work. Investigations of the problems of contemporary research have elicited a number of responses, which largely reduce to three categories: ethics training, increased surveillance, and reorganization of the scientific enterprise. As noted, many commentators have identified important aspects of the contemporary crisis of science, but since they have not traced them to their roots in the entrepreneurial ideal, they are unable to offer a response adequate to the problem. These failures occur in one of two ways: either these responses avoid addressing the structural factors at play in the crisis of science or they improperly diagnose what those structural problems are. These diagnostic failures lead to responses that will not solve the problem.

As an example of the first diagnostic failure, take the 2017 National Academy of Sciences report on research integrity. The report, like many other commentators, recognizes that a major contributor to the reproducibility crisis is the competitive nature of today's research.[78] However, it considers this problem as outside the reach of ethics or integrity, stating that "addressing larger structural issues in US research funding and organization is beyond the scope of this study."[79] Structures are divorced from ethics. When one avoids these features of science, and more importantly the ways they become embedded in the character of the researcher, one is left with only two responses: education or surveillance.

Responsible conduct of research education has long been a stock response to problems of scientific misconduct, and the National Academy of Sciences turned to it again as one of its major solutions to the current crises of fraud and reproducibility.[80] This answer seems simple enough:

[77] E.g., China Scherz, *Having People, Having Heart* (Chicago: University of Chicago Press, 2014).
[78] National Academies of Sciences, Engineering, and Medicine, *Fostering Integrity in Research*, 98–103.
[79] National Academies of Sciences, Engineering, and Medicine, *Fostering Integrity in Research*, 208.
[80] National Academies of Sciences, Engineering, and Medicine, *Fostering Integrity in Research*, 195–204.

if one wants to stop research misconduct, then one should tell people what misconduct is, that it is wrong, and that they should not do it. Researchers have been forced to undergo this kind of training for thirty years now, in their graduate training, in seminars, and now in online form. Yet the crisis in science has gotten worse rather than better as this education program has proceeded. The problem is that this framework depends on a poor model of moral formation. No matter how advanced these educational modules are, they will not affect the role of everyday life in shaping character. Such a theory of moral formation thinks that a mere couple of hours of propositional learning based in case studies will transform researchers when everything in their daily practice of research pushes against these guidelines. It does not investigate what researchers' character traits are and how these are being formed. It is for this reason that an adequate model of the moral formation of scientists and the moral ideal into which they are being shaped is essential to any response to the crisis of science. Chapters 3 and 4 will provide a better model of that formation through an investigation into the role of practices in shaping character in contemporary science.

A second alternative to standard forms of contemporary education through online courses is to put a greater focus on mechanisms of surveillance, monitoring, evaluation, and audit. If one fears that researchers are dishonest, then one can consistently check up on their work. Bodies are increasingly creating formal structures of rules and mechanisms for ensuring that researchers adhere to those rules. If journal editors fear image manipulation in the figures of articles, then they can use software that detects image manipulation to analyze all submissions. If they fear plagiarism, they can use plagiarism detection software. They can require that all authors check their cell lines. To enable the replication of the data analysis, they can demand that all data be made public. To fight publication management, they can require a precise delineation of each author's contribution to the project. If they fear bias, they can require conflict of interest statements. Many major journals have implemented some or all of these precautions.

These are all reasonable steps that should be implemented. Unfortunately, they alone will not solve the reproducibility crisis because they fail to recognize how much the scientific enterprise depends on the honesty and character of the individual researcher. In regard to fraud, most of these surveillance mechanisms examine the end product: whether an image in the final published figure was manipulated or whether the final statistical analysis was performed correctly on the total data set. Yet many of the most egregious cases of fraud involve the fabrication of the data sets themselves. As noted above, Diederik Stapel did not perform any experiments. He just made up the experimental data sets to fit his conclusions. While few researchers would go that far, there are many points at which a researcher could manipulate the final result by choosing which data go into the data set, throwing out outliers, failing to record disconfirming instances, substituting the results of a completely different experiment. While these cases of fraud can be detected through sophisticated statistical analyses and through examining laboratory notebooks, these are difficult, time-consuming, and expensive investigations, which are generally undertaken only when major fraud becomes apparent. The fear is that the simple tools of mass surveillance will only push the fraud back one level.

Similar problems arise in relation to the other aspects of the crisis of science. These mechanisms simply would not catch sloppy or careless research. Poor laboratory practices would still occur. Detailed protocols and complete data sets could possibly aid replication, but the problem is that most research simply is not replicated because it is expensive, time-consuming, and there is little incentive, either personal or professional, to engage in extensive replication of others' experiments. Only in a few influential studies would anyone even undertake replication. As will be discussed later, much of the knowledge of how to do an experiment is tacit, and therefore will not make it into a published guide. Further, analyzing data requires many small decisions, not all of which will be recorded, so these published data sets would not be sufficient on their own to allow replication in many cases. Finally, even if conflicts of interest are reported, it does not mean that anyone will take them into account when reading the paper. In many fields of important biomedical research, almost all the major figures have conflicts of interest in the form of a stake in the company or a consulting deal with a company, so one could not ignore such research without casting aside most of the research in the field.[81] While these audit mechanisms for fraud, reproducibility, and conflict of interest are not a problem in themselves, we should not expect them to solve the crisis in science.

On the other hand, some of the surveillance tools developed to enhance innovation are major threats to contemporary research since they reinforce some of the worst aspects of entrepreneurial science. Recognizing the

[81] Jeremy A. Greene, *Prescribing by Numbers: Drugs and the Definition of Disease* (Baltimore, MD: Johns Hopkins University Press, 2008); Carl Elliott, *White Coat, Black Hat: Adventures on the Dark Side of Medicine* (Boston: Beacon Press, 2011).

problem of falling rates of innovation and the need for innovation in an entrepreneurial research culture, many universities and funders have attempted to develop tools to measure the innovative output of scholars. Since there is no good way to quantitatively measure what is high-quality, innovative research, these institutions are forced to rely on secondary, measurable substitutes for innovation. Examples of this include considering the number of publications as a measure of productivity, the importance of the journal as a measure of a paper's quality, the number of citations as a measure of scholarly impact, or the mentions on social media as a measure of social impact. These developments have given rise to a whole field called bibliometrics that attempts to quantitatively analyze research outputs. Institutions use these numerical measures to compare researchers in making hiring, funding, and promotion decisions.

A number of researchers in this field are warning that the use of these metrics has gone too far. In 2015, a number of science policy experts issued the Leiden Manifesto, arguing, "We risk damaging the system with the very tools designed to improve it, as evaluation is increasingly implemented by organizations without knowledge of, or advice on, good practice and interpretation."[82] Similarly, a recent review of the United Kingdom's Research Excellence Framework, the evaluation system of the research productivity of the universities that is used in budgeting research funds, acknowledged that many of the tools used to analyze research productivity can skew research programs and hiring decisions in problematic ways, a nod to the many critiques of the audit culture in UK universities.[83] The bibliometric experts offer a number of rules for using these metrics well, but their basic point is that these tools should assist, rather than replace, the expert judgment of the reviewers. Too often, however, reviewers use the metrics to replace human judgment, in line with the more widespread replacement of prudential judgment with quantitative analytics like cost–benefit analysis in public policy.[84]

[82] Diana Hicks et al., "Bibliometrics: The Leiden Manifesto for Research Metrics," *Nature News* 520, no. 7548 (April 23, 2015): 429. See also the Declaration on Research Assessment (DORA), first written under the auspices of American Society for Cell Biology, but endorsed by many other stakeholders in science, at https://sfdora.org.

[83] Nicholas Stern, "Building on Success and Learning from Experience: An Independent Review of the Research Excellence Framework" (London: Department for Business, Energy & Industrial Strategy, 2016). For discussions of audit, see Michael Power, *The Audit Society: Rituals of Verification* (Oxford: Oxford University Press, 1999); Marilyn Strathern, *Audit Cultures: Anthropological Studies in Accountability, Ethics and the Academy* (New York: Routledge, 2000).

[84] For a general analysis of this tendency in modern public policy of using quantitative metrics as a seemingly neutral, transparent, more democratic way to replace the possible bias of elite expertise, see Theodore Porter, *Trust in Numbers* (Princeton, NJ: Princeton University Press, 1996).

The problem is that these metrics that attempt to gauge the level of innovation, which is difficult to quantify, using other, more measurable outputs introduce their own notion of what is a valuable outcome of the research process and therefore skew the research enterprise to meet those outcome targets. If the target metric is number of publications, then researchers will aim to produce as many publications as possible, which will generally end up with more publications in lower-tier journals, a focus on safe projects, and less time spent reproducing and validating experiments in other systems. If an outcome is social media exposure as measured by an altmetric, then the researcher will spend more time on social media and less in the laboratory. These aspects undermine innovation and reproducibility. Further, in their attempt to meet these targets, researchers are tempted to commit various kinds of misconduct. To get a higher number of publications, researchers publish with so-called predatory publishers, journals that require a payment for publication but have little peer review. Or there are scandals over self-peer review in which authors give editors fake addresses for recommended peer reviewers so that they can approve their own work. If the requirement is to place a paper in a journal with a high impact factor, meaning one that is cited frequently, then researchers will be tempted to commit fraud to succeed in publishing. If the metric is number of citations, then researchers will be tempted to set up citation circles whereby groups of researchers go out of their way to cite each other or to cite a certain journal. More importantly, all of these measures reinforce a competitive, short-term focus in research.

Even when commentators try to take into account the broader structural features of the contemporary problems of science, they frequently fail to trace these issues to the correct source: the encouragement of entrepreneurial characteristics in researchers. For example, a group of senior biomedical researchers recognize problems of hypercompetition and burnout but trace these problems to the idea that research funds would expand indefinitely. They encourage a managed stabilization of federal research funding alongside a greater focus on originality in the evaluation of grants.[85] These are fine suggestions, but they do not truly address the entrepreneurial focus on competition.

Daniel Sarewitz, in a widely discussed article, took an entirely different tack and placed the blame on curiosity-driven research.[86] He does not

[85] Alberts et al., "Rescuing US Biomedical Research from Its Systemic Flaws."
[86] Daniel Sarewitz, "Saving Science," *New Atlantis*, no. 49 (Spring/Summer 2016): 4–40.

think scientific research devotes as much attention to actual problems in the world, nor is it disciplined by the requirement to generate products. Instead, he wants more public and government control over research, hearkening back to the days of military control over research such as in the Manhattan Project, other technological developments in World War II, and the Advanced Research Projects Agency (ARPA). The government and patient advocates should oversee research to ensure that it is developing into concrete applications. In some ways, his claims are just another step in a long-term struggle over the role of science – whether it is a producer of technologies that should be controlled through bureaucratic management or a practice dedicated to the good of knowledge that depends on individual researchers. This struggle can be seen in the arguments of the 1930s between J. D. Bernal and Michael Polanyi over the organization of British research and in Cold War comparisons between planned Soviet science and investigator-driven US science.[87] Ironically for Sarewitz's argument, it was precisely the US military that encouraged pure science after the war instead of taking full control of research planning.[88] They recognized that planning worked well for short-term engineering projects but not for the long-term acquisition of knowledge.

Indeed, such a solution would reinforce the neoliberal focus on using science to drive economic growth and its resulting problem of investigator burnout. In a 2015 interview with an editor of the British biological journal *Development*, the geneticist Didier Stainier explained why he relocated his laboratory from the United States to the Max Planck Institute in Germany:

> In the USA, and in other countries, the emphasis – and hence the funding – is increasingly being placed on translational research ... In order to keep running a lab in the USA investigating several different topics, I would have had to move to more translational work. I had a hard time imagining myself not being able to pursue the kind of basic research required for innovative translational work.[89]

Already, as we have seen, government policy has helped create a research culture that focuses on the short-term production of tools rather than the long-term development of understanding.

[87] J. D. Bernal, *The Social Function of Science* (London: Faber & Faber, 2010); Michael Polanyi, "The Republic of Science: Its Political and Economic Theory," *Minerva* 38, no. 1 (2000): 1–21.

[88] Mirowski, *Science-Mart*, 105.

[89] Seema Grewal, "An Interview with Didier Stainier," *Development* 142, no. 17 (September 1, 2015): 2861–3.

Federal funding is not in itself a threat to fundamental research. Even astute critics such as Polanyi who are skeptical of government's role in the management of science see a role for the state in subsidizing research. The potential problems lie in the way in which government (or any other funder) influences the aims and practices of science, perhaps most perniciously through a growing fondness for short-term funding initiatives responding to the hype of new technologies. President Barack Obama, for instance, proposed the BRAIN Initiative (2013), the Precision Medicine Initiative (2015), and the Cancer Moonshot Task Force (2016). But he is only the latest in a long line of presidents who have proposed big projects of this type, including stem cell research, the Human Genome Project, and even the original War on Cancer of the 1960s and 1970s. Such initiatives typically encourage researchers to look for immediate technological solutions and fixes, not deeper understandings of underlying processes and mechanisms.

The problems of this approach are compounded by the inconsistency of government funding practices. From time to time, science funding dramatically surges, as in the doubling of the National Institutes of Health budget from 1998 to 2003 or the research investment in the economic stimulus package of 2009. Such expansions typically lead universities to train more students, start new laboratories, and develop new centers. Then real investment in research suddenly decreases, and the laboratories and centers are left scrambling to find grants, while students desperately try to track down new jobs.

Since funding is aimed at specific initiatives rather than a broad portfolio of basic research, researchers are left with no choice but to respond to government incentives and enter those areas tied to these initiatives, a process that engenders chronic cynicism toward grant writing as young faculty try to describe why their investigations of the cell cycle or embryonic development are essential to curing cancer. Such a model of research funding offers no security, leaving scientists demoralized and many of the best young researchers seeking other careers.

Even when scientists obtain government grants, they are all but discouraged from engaging in deep theoretical exploration. With a few exceptions, government grants fund only two to five years of research and oblige the investigator to focus on the straightforward analysis of a hypothesis that is almost predictably confirmable. A project that is outside the mainstream or that is overly ambitious faces difficulties in receiving approval from a grant review panel, especially if it is proposed by a young scientist at the peak of her creativity.

Scientists who rely on government grants must continually calibrate their research interests to the changing funding regimes. Usually this requires that they design projects with near-term goals promising immediate practical benefits. Recent congressional debates over National Science Foundation funding resulted in the requirement that all grants show a clear payoff for the national interest. Many fear this will lead only to an increased emphasis on economic competitiveness over real scientific progress.[90] Thus, an incorrect understanding of the causes of the problems of science can lead to improper solutions.

Conclusion

If we are to address the contemporary problems in scientific research, it is imperative that we address the problems raised by the moral exemplar of the scientist entrepreneur: competitiveness, market responsiveness, and seeing knowledge and nature as potential property, among others. One could begin to do this by changing laws, thus changing structures, and hopefully changing character. Yet, as Wendy Brown cautions in relation to political processes, even if almost revolutionary changes in policy were possible, such an approach might not be successful:

> [S]trong bank regulation . . ., public reinvestment in education, . . . or even wealth redistribution, for example, could coexist with the economization of political life, the remaking of education by business metrics, or the formulation of elections as marketplaces . . . Thus, neoliberal economic policy could be paused or reversed while the deleterious effects of neoliberal reason on democracy continued apace unless replaced with another order of political and social reason.[91]

What is needed is a transformation of how one thinks, how one sees the world, and one's habits and dispositions. To address these problems, it is important to have a proper model of how one is formed in the character traits of the scientist entrepreneur, as well as how one can form other kinds of habits and dispositions.

If changes to laws and regulations alone will not be enough to fix the problems of science, then perhaps we should turn to the question of how we can address the problems in the character of the scientist. If the problem lies in the habits and dispositions of the moral ideal of

[90] Paul Basken, "U.S. House Backs New Bid to Require 'National Interest' Certification for NSF Grants," *The Chronicle of Higher Education*, February 11, 2016.

[91] Brown, *Undoing the Demos*, 201–2.

the entrepreneur that scientists are trained to emulate, then we ought to ask how we can shape those habits and dispositions in a different way. One of the most productive recent paradigms for addressing character formation in philosophy, social theory, and Christian ethics has focused on the role of practices. In Christian and philosophical ethics, virtue ethics has emphasized the role of social practices in shaping habits and disposition. By engaging in communal practices directed toward certain goods, one begins to form habits that become solidified over time. However, this is not the only way of theorizing the effects of practices on character. A second major paradigm of practice theory emerging from French social theory has influenced social scientific analyses of ethics. While this model also looks at how practices relate to social institutions, it also brings into play practices of the self, techniques used by the individual to shape her own character even in situations where the social institutions may be challenging. Chapters 3 and 4 will examine what these different models of practice reveal about the current practice of science and how they might be used to address its problems. These discussions will lead in Chapter 5 to a comparison of how effective these models might be as responses to these problems.

CHAPTER 3

Teleology and the Craft of Science

The last chapter described three vices embedded in the ideal of the contemporary scientist: competitiveness, market responsiveness, and reductively seeing the things of the world as potential property. While competitiveness is clearly connected to the most blatant problems of contemporary research such as fraud and unreproducible results, this chapter will argue that the vision of the world as a set of commodifiable objects is actually an older and more deeply ingrained flaw in the ideal character of the scientist. In fact, it is the necessary precondition for the other traits of the scientist entrepreneur. Unless one can transform knowledge into marketable intellectual property, one cannot engage in a market and thus cannot be responsive to the market. Unless one is engaged in a market, one cannot engage in a competitive struggle for money and market position. This vision preceded the neoliberal structures of science and enables them.

The importance of an instrumental relationship to the world to entrepreneurial science draws attention to a larger stream of critique. Many commentators, especially those influenced by Aristotle, have long argued that one of the central problems of modernity is the loss of an understanding of teleology. Things in the world no longer have intrinsic purposes and ends, and hence seem to lose any inherent meaning or value. Without a vision of inherent purposes embedded in the world, one is constrained to relate to the world only in an instrumental manner. Christian theologians have drawn extensively on this vein of critique especially in relation to the environmental crisis. If the things around us are no more than items for our use, then it seems to lead to the conclusion that we can degrade the environment at will. While some commentators, such as Lynn White, see Christianity itself as to blame for these problems, other theologians have shown that such an instrumental stance is opposed to biblical and

Christian conceptions of the world.[1] In *The Environment and Christian Ethics*, Michael Northcott provides one of the most thorough defenses of this latter view, arguing that the Hebrew Bible shows "that moral values and purposes were enshrined in the nature of created order."[2] In the biblical text, God creates an orderly universe, with meaning and purpose, and calls it good. According to Northcott, this understanding of creation persisted throughout the theological tradition in the concept of natural law.[3] More recently, in his encyclical *Laudato si'*, Pope Francis made a similar point, arguing that "our insistence that each human being is an image of God should not make us overlook the fact that each creature has its own purpose."[4] Thus, viewing the world in a merely instrumental manner is a mistake from the Christian perspective.

More importantly for the argument of this book, the loss of a teleological vision leads to a loss of an understanding that there might be purposes in human life. Without understanding humans as having certain ends to pursue – purposes whose achievement can lead to a flourishing life or *eudaimonia* – ethics begins to become an arbitrary list of duties disconnected from any end. This argument was influentially stated by Elizabeth Anscombe, but was most fully fleshed out by Alasdair MacIntyre, who influenced later generations of Aristotelian ethicists.[5] Importantly, one needs to be a certain kind of person in order to flourish. Thus, when ethicists lose a teleological understanding of the world and human life, they cease to focus on character. Instead, they merely argue for rules for specific action. According to this Aristotelian critique, without an understanding of teleology, one's character becomes incoherent because there

[1] Lynn White, "The Historical Roots of Our Ecologic Crisis," in *Philosophy and Technology*, ed. Carl Mitcham and Robert Mackey (New York: Free Press, 1983), 259–65.

[2] Michael S. Northcott, *The Environment and Christian Ethics* (New York: Cambridge University Press, 1996), 164.

[3] Northcott, *The Environment and Christian Ethics*, 199–256.

[4] Francis, *Laudato si'*, 2015, 84, w2.vatican.va/content/francesco/en/encyclicals/documents/papa-francesco_20150524_enciclica-laudato-si.html. For further discussion of the use of science in this encyclical, see Paul Scherz, "Laudato Si' and the Use of Scientific Research in Theology and Public Policy," *The Heythrop Journal*, May 1, 2016, doi.org/10.1111/heyj.12337. For problems with the idea of nature in relation to scientific manipulation, see, among others, Gerald McKenny, *Biotechnology, Human Nature, and Christian Ethics* (New York: Cambridge University Press, 2018); Kathryn Tanner, "Grace without Nature," in *Without Nature?*, ed. David Albertson and Cabell King (New York: Fordham University Press, 2010), 363–76. Neither of these last two authors would affirm a purely instrumentalist approach to nature, though.

[5] G. E. M. Anscombe, "Modern Moral Philosophy," *Philosophy* 33, no. 124 (1958): 1–16; Alasdair MacIntyre, *After Virtue: A Study in Moral Theory*, 2nd edition (Notre Dame, IN: University of Notre Dame Press, 1984). For a broader review of the history of virtue ethics in theology, see David Cloutier and William Mattison III, "The Resurgence of Virtue in Recent Moral Theology," *Journal of Moral Theology* 3, no. 1 (2014): 228–59.

is no end that one is trying to pursue. Further, one's actions become governed by technique, meaning that one acts instrumentally so as to achieve whatever goods one desires rather than things and characteristics that are objectively good. It is because of this that the mainstream response to the current crisis of science has focused not on character traits and moral formation but on integrity training and surveillance. These methods are simply the training for and evaluation of rule-following behavior.

The question then becomes how an individual can regain a teleological vision of life that would shape a character aimed at flourishing. Many virtue ethicists look to the moral formation of premodern individuals, whose desires arose from virtues shaped by communal practices in line with an integrated, teleological system of value. These elements capture some of what moral formation is in contemporary life. Christian ethicists have looked to the community of the parish or, more recently, to the practices of community organizing to provide such moral formation.[6] The most well-developed model of this kind of character formation is that of MacIntyre's focus on *techne*. MacIntyre's description of *techne* hearkens back to a vision of premodern craft practice, a focus that reflects the importance of the model of craft in Aristotle's understanding of ethics.[7] A *techne* combines a pursuit of the goods of excellence that are internal to a craft with a moral education in the virtues of character necessary for achieving such goods. This education occurs through an apprenticeship dependent on the authority of a master of the craft. The work of Michael Polanyi and Thomas Kuhn demonstrate that such a craft model of character formation works well for science, in which the trainee undergoes an apprenticeship through which her worldview is shaped by older scientists and which depends on tacit learning through embodied practices with material. MacIntyre himself recognizes that natural science is a good model for his understanding of tradition and moral formation. However, there are points at which scientific training departs from MacIntyre's program.

This chapter will therefore examine how well this model of moral education through social practices fits scientific training. Drawing on the work of the sociologist Max Weber, it will first give a more thorough description of the problems of the modern worldview and its associated character, especially as it applies to natural science. Then, it will turn to

[6] Stanley Hauerwas, *A Community of Character: Toward a Constructive Christian Social Ethic* (Notre Dame, IN: University of Notre Dame Press, 1991); Jeffrey Stout, *Blessed Are the Organized: Grassroots Democracy in America* (Princeton, NJ: Princeton University Press, 2010).

[7] Tom Angier, *Techne in Aristotle's Ethics: Crafting the Moral Life* (London: Bloomsbury Academic, 2012).

the Aristotelian solution to these problems through an apprenticeship in communal practices, showing the ways in which scientific training fits this model quite well. The second half of the chapter will turn to the problems of this model, arguing that the daily practices that one engages in the laboratory form one in a reductionist worldview. This issue leads to a need to consider other models of moral formation.

The Weberian Scientific Ideal

Written just over a hundred years ago in 1917, Weber's essay "Science as a Vocation" has been tremendously influential.[8] Crystallizing many of the insights of his larger work, it has provided a description of a moral ideal for intellectual work amid the problems of modernity. There are of course many criticisms of his work, such as whether his idea of disenchantment actually holds, or even what it means.[9] There are normative critiques of whether one should behave in a Weberian manner, especially regarding how politics relates to one's teaching and intellectual work.[10] Even with these problems, though, his overview of how intellectuals function in society and his characterization of the modern individual are tremendously influential. Many find that his scientific ideals still have much to teach us.[11] Even those who disagree with him admit that "the contemporary vision of the world ... is predominantly, although not perhaps always in detail, Weberian."[12] Thus a discussion of Weber provides a good starting point to serve as a foil for the Aristotelian response to the moral problems of modernity. Three elements of Weber's diagnosis of the problems of modernity are especially helpful: the separation of value spheres, the fact–value distinction, and the dominance of technique.

Central to Weber's understanding of modern problems is the loss of a unifying framework for understanding the world. Natural sciences "are apt to make the belief that there is such a thing as the 'meaning' of the universe die out at its very roots."[13] This creates problems because there are many

[8] Max Weber, "Science as a Vocation," in *From Max Weber*, ed. H. H. Gerth and C. Wright Mills (New York: Oxford University Press, 1958), 129–59.

[9] E.g., Jason A. Josephson-Storm, *The Myth of Disenchantment: Magic, Modernity, and the Birth of the Human Sciences* (Chicago: University of Chicago Press, 2017).

[10] For example, see the anthropological debates in the June 1995 issues of *Current Anthropology.*

[11] Chad Wellmon, "Whatever Happened to General Education?," *Hedgehog Review* 19, no. 1 (2017). See also the related discussion in Chapter 5.

[12] MacIntyre, *After Virtue,* 101.

[13] Weber, "Science as a Vocation," 142.

genuine goods that one could pursue: religion, family, art, science, politics, etc. With the loss of a unified vision of the good and social changes such as the separation of work and home, different goods started to become independent of each other and come into rationally irresolvable conflict. This situation is what Weber calls the separation of value spheres: "The various value spheres of the world stand in irreconcilable conflict with each other."[14] The good of work and career clashes with that of family. Weber terms these problems a revival of polytheism, with each area of life having its own god, its own hold on the person: "Many old gods ascend from the graves; they are disenchanted and hence take the form of impersonal forces. They strive to gain power over our lives and again they resume their eternal struggle with one another."[15] A central question for ethics, the central question for a eudaimonistic system, is how to order these goods, to decide which to prefer over others in a particular situation and in an overall life plan. While to some extent these decisions will always be personal, older traditions and social structures provided a hierarchical ordering of goods and practices that assisted personal and political choice.[16] For Weber, one must make an existential choice as to the sphere of life that will command one's ultimate allegiance: "[I]t is necessary to make a decisive choice."[17] In Weber's view, the scientist can devote herself solely to the good of scientific truth, but such devotion requires a rejection of faith and politics. Each person must find and obey "the demon who holds the fibers of his very life."[18] Such choices, however, lead to difficulties in living as an integrated whole, as one must reject some areas of one's existence.

These goods and ends that inhabit the different spheres have become mere values, values that are made ends by arbitrary subjective choice. Yet modernity still retains the Enlightenment confidence in science. Reason is not done away with; it merely shifts its considerations from objective ends to the means to reach subjectively chosen ends. Reason helps one to gain clarity on what means are necessary and forces one to decide whether these are acceptable. Science can still discover the facts about the world that condition our action and create the technologies that help us succeed in action. Thus is born what in Weber is the distinction between facts and

14 Weber, "Science as a Vocation," 147.
15 Weber, "Science as a Vocation," 149.
16 Alasdair MacIntyre, *Whose Justice? Which Rationality?* (Notre Dame, IN: University of Notre Dame Press, 1988), 33–4.
17 Weber, "Science as a Vocation," 152.
18 Weber, "Science as a Vocation," 156.

values.[19] Objective science provides facts that help individuals and society to choose the means for obtaining their ends. Scientists show integrity especially when they force people to face "facts that are inconvenient for their party opinions."[20] Hence, the research university, with its disciplinary fragmentation and its lack of moral education, is justified so long as its natural and social sciences provide facts to aid the management of society and scientists pursue that knowledge with integrity.

Philosophers have pointed to problems with this distinction between objective, atomistic facts and subjective values. It is based on an understanding of science challenged by Pierre Duhem and W. V. O. Quine's demonstrations that all observation is theory laden.[21] Every fact is an interpretation of an experiment or an experience through the lens of whatever scientific theory one holds. Because of these different interpretations of reality, arguments between different theoretical viewpoints cannot be solved by a straightforward appeal to evidence. These debates require arguments between whole scientific paradigms. The reason the fact–value distinction has gained such a hold on our society is because it is really derived from claims from moral philosophy, the moral claim that one cannot derive "ought" from "is."[22] First put forward by Hume, and developed by later ethicists, it is the claim that observed facts (what is) cannot lead to moral conclusions (what one ought to do) because they are claims of a different kind.[23] Such a position makes a natural law or Aristotelian theory impossible. In turn, Aristotelians have argued that this position holds only if one has a wholly mechanistic vision of the universe like that of early modern science, but not if one accepts teleology.[24]

This loss of a basic connection between facts and values means that knowledge divorced from the ability to reveal true good in the world becomes purely instrumental. Facts do not help one to choose ends, because the ends are determined by one's commitment to a certain value sphere or political project that is chosen largely for nonrational reasons.

[19] Weber, "Science as a Vocation," 146.

[20] Weber, "Science as a Vocation," 147.

[21] Pierre Duhem, *The Aim and Structure of Physical Theory* (Princeton, NJ: Princeton University Press, 1954); W. V. Quine, "Two Dogmas of Empiricism," in *From a Logical Point of View: Nine Logico-Philosophical Essays*, 2nd revised edition (Cambridge, MA: Harvard University Press, 1980), 20–47.

[22] MacIntyre, *After Virtue*, 56–9.

[23] David Hume, *A Treatise of Human Nature: A Critical Edition*, ed. David Fate Norton and Mary J. Norton (New York: Clarendon Press, 2011); G. E. Moore, *Principia Ethica* (Cambridge: Cambridge University Press, 1922); John Finnis, *Natural Law and Natural Rights*, 2nd edition (New York: Oxford University Press, 2011), 33–58.

[24] MacIntyre, *After Virtue*, 148; Philippa Foot, *Natural Goodness* (Oxford: Clarendon Press, 2003).

Instead, facts help one to achieve ends that are set for other reasons. This conception licenses certain modern notions of expertise, especially, for our discussion, bureaucratic managerial expertise. Knowledge fundamentally gives one control over technique.

Technique is a rational means to achieve an end with measurable effectiveness.[25] This focus on measuring and calculation is essential to Weber's conception of modernity, especially in his concept of disenchantment. In Weber's view, modern rationalization "means that principally there are no mysterious incalculable forces that come into play, but rather that one can, in principle, master all things by calculation. This means that the world is disenchanted."[26] Disenchantment is not primarily about religion, although it does have a bearing on religious understandings of divine action in the world, but is really a change to a purely mathematical understanding of causality in the world.

Here, we see many aspects of contemporary entrepreneurial science set in play. There is no fundamental meaning given to knowledge or things in the world, which opens them up to commodification. Since knowledge is not intrinsically meaningful, its primary purpose is at the service of technique, becoming a purely instrumental good aimed at achieving external organizational or personal goals. All that is missing is the role of the market as the prime technique for attaining knowledge and achieving goals. Once that is put into place, one has entrepreneurial science.

Aristotelian Apprenticeship

As noted in the preceding sections, Aristotelian and many Christian commentators disagree with all three aspects of the Weberian description of the contemporary individual as a moral ideal. Facts and values are not separate because things in the world have inherent purposes, and nature is considered fundamentally good. Since there are true goods, one must use knowledge for more than mere technical manipulation. Humans, as beings with purposes, have a set of ends that help them to integrate the different areas of their lives, as difficult as that may be in practice in many cases.

[25] MacIntyre, *After Virtue*, 30. This analysis is connected to broader criticisms of technology in modernity, e.g., Jacques Ellul's definition of technique: "technique is the totality of methods rationally arrived at and having absolute efficacy (for a given stage of development) in every field of human activity." Jacques Ellul, *The Technological Society*, trans. John Wilkinson (New York: Vintage Books, 1964), xxv.

[26] Weber, "Science as a Vocation," 139.

Such a eudaimonistic understanding of nature and of the person presents a very different model of what a moral life would be.

Yet, given that many people think of the world and human life starting from Weberian or neoliberal principles and that our institutions shape people in those worldviews, how can one be formed in the Aristotelian ideal? MacIntyre and others have answered that question by arguing that one must take on a teleological mode of understanding the world and develop an appreciation for the importance of character. Recent scholars have settled on the notion of social practices as the most important way of shaping worldview and character. One of the key analogies for understanding such a process of moral education through practices is that of craft practice, what the Greeks called *techne*. Such craft practices unite teleology, education, authority, and a commitment to realism. This section will first examine MacIntyre's general model of moral formation through crafts, before turning to discuss how Michael Polanyi and, in a way, Thomas Kuhn applied very similar models to moral formation in science. It turns out that an Aristotelian model of communal formation converges well with recent sophisticated analyses of scientific training.

Social Practices and Moral Formation

The teleology inherent in the tools and goals of crafts stands opposed to the contemporary distinction between fact and value, is and ought. Tools are obvious examples of purposiveness. The normative status of a hammer is dependent on its actual ability to drive in nails. If it fails to drive in nails well, then it is a bad hammer. If it suffices for that task, then it is a good hammer. Following Plato and Aristotle, all *technai*, like all tools, exist for the purpose of achieving some good,[27] although, like Aristotle, MacIntyre argues that "the skills of a *techne* may be exercised with the purpose of achieving what is not in fact a good."[28] The use of the term *techne* also allows him to focus on practices that manipulate the world.[29] Since such practices require at least some sort of realism, they preclude unrestrained skepticism, relativism, or perspectivism.

[27] MacIntyre, *Whose Justice? Which Rationality?*, 72–3; Alasdair MacIntyre, *Three Rival Versions of Moral Enquiry: Encyclopaedia, Genealogy, and Tradition* (Notre Dame, IN: University of Notre Dame Press, 1990), 61.

[28] MacIntyre, *Three Rival Versions of Moral Enquiry*, 62.

[29] MacIntyre, *Three Rival Versions of Moral Enquiry*, 68.

The goods sought by a *techne* are distinct from those sought by technique. There are goods that everyone needs in order to accomplish any task whatsoever, such as food, shelter, or health, as well as goods that serve as rewards for successful performance, such as wealth and honor.[30] These goods are different from those internal to the different spheres of life, such as love in the family or knowledge for the sciences, because they are applicable to all spheres, and thus MacIntyre calls them external goods. In contrast, deep engagement in a *techne* drives the pursuit of goods that are only possible to achieve through excellence in that activity. For example, while a chess player can achieve fame by victory, she would not seek this achievement through cheating, because that would defeat the achievement gained through excellent play.[31] This excellence at play is a good internal to the practice of chess. Similarly, the scientist can perform research merely as a job or even to become spectacularly wealthy by starting a company, yet these are goods external to the practice of science itself. The practice of science is aimed at the solution of the puzzles that nature presents to us, at gaining a better insight into the nature of the reality that surrounds us, and, ultimately, at attaining truth. Therefore, somewhat clear distinctions can be made between internal and external goods.

In actual fact, however, the relations between these two types of goods are much more complicated. External goods are necessary for the achievement of goods internal to a practice but also threaten to undermine them.[32] For example, publications are how a scientist relates his findings to the community and are thus essential to one of the goods internal to science: the sharing of knowledge. At the same time, as we have seen, publications are also a route to tenure and grants, so publications both mark the achievement of a good internal to the practice of science and are used as a metric of achievement to choose who will have access to the goods external to science, goods necessary for continuing the very practice of science. As we have seen in Chapter 1, scientists can commit fraud or perform sloppy research to attain only the external goods, thus undermining the practice.

This tension between internal and external goods is reflected in a tension between a *techne* and its institutional embodiment. Every practice needs to be supported by an institution that provides access to external

[30] MacIntyre, *After Virtue*, 188.
[31] MacIntyre, *After Virtue*, 188.
[32] MacIntyre, *Whose Justice? Which Rationality?*, 35.

goods. Biology needs research institutes and biotechnology companies; medicine needs hospitals and insurance companies; and art needs galleries and museums. However, practice and institution can be conceptually separated and must be for a proper ethical analysis. Institutions will pursue the external goods necessary for its attendant *techne*, thus threatening *technai* by subordinating their ends to bureaucratic imperatives. One can oppose illegitimate expansions of technique through an appeal to the internal goods of the practice and to the virtues. "Without [the virtues] ... practices could not resist the corrupting power of institutions."[33] Thus the idea of *techne*, through its focus on goods internal to practice, is a central tool for opposing the fact–value distinction and containing technical expertise.

Where MacIntyre expands beyond Aristotle is in his focus on moral formation in the practice of *techne*, an analysis that combines Aristotle's practical wisdom with productive activity. To engage in a craft, one must first undergo an apprenticeship and make oneself into a particular kind of person, submitting to the knowledge and authority of a master to learn "what intellectual and moral habits it is which we must cultivate if we are to become effective self-moved participants."[34] These intellectual and moral habits that help one to achieve goods internal to a practice are the virtues.[35] At the first stage, the apprentice must accept the maxims of the practice that the teacher gives on faith, but as one practices the craft following the example of the master, one begins to understand the reason behind the rules and how they aid in successful accomplishment of the craft.[36] The apprentice starts to develop the excellence of the craft and its virtues, learning when rules should be strictly applied and when they can be set aside for further achievement.[37] Here, we see the analogy between technical and moral education. As a child, one begins by learning mere rules, but as one grows and follows the rules, one develops the virtues and experience that give a deeper understanding of the moral life. Practical wisdom allows a more accurate application of moral maxims. Yet this moral autonomy first requires a submission to authority.[38] As James

[33] MacIntyre, *After Virtue*, 194.
[34] MacIntyre, *Three Rival Versions of Moral Enquiry*, 63.
[35] MacIntyre, *After Virtue*, 191.
[36] MacIntyre, *Whose Justice? Which Rationality?*, 115.
[37] MacIntyre, *Whose Justice? Which Rationality?*, 31.
[38] There was a growing emphasis on authority in MacIntyre's works as discussed in Jean Porter, "Tradition in the Recent Work of Alasdair MacIntyre," in *Alasdair MacIntyre*, ed. Mark Murphy (New York: Cambridge University Press, 2003), 38–69.

Mackey notes in his *Power and Christian Ethics*, the kind of power used in an apprenticeship is a beneficial power, one aimed at the good of the individual on whom the power operates because it aims at shaping her into a self-moving, full member of a community.[39]

Education into a *techne* is thus a process of self-development over time. Similarly, every *techne* is itself in the process of historical development. Those who engage in the practice are constantly extending its aims and altering its standards of excellence.[40] For example, painting changed from an unrepresentative icon, to emphasizing increased representational accuracy, to using the naturalistic portrait as an icon.[41] After education into traditional techniques, a painter like Rembrandt transforms the aims and standards of the practice. Thus, a *techne* becomes a historical tradition shaped by its practitioners. Yet these revolutions are possible only because the practitioners first submitted to authority.

Apprenticeship in Science

This conception of moral formation through *techne* shares important characteristics with the views of science developed by Michael Polanyi and Thomas Kuhn.[42] In their work, the process of education by authority based in a specific community is central. Polanyi sees science not as an abstract body of knowledge but as embedded in a community of intellectual elites who maintain the tradition of science through an existential commitment to forms of personal knowledge. Polanyi recognizes that no one accepts objective, impersonal facts separate from values.[43] Scientific

[39] James Mackey, *Power and Christian Ethics* (New York: Cambridge University Press, 1994), 98.

[40] MacIntyre, *Three Rival Versions of Moral Enquiry*, 66.

[41] MacIntyre, *After Virtue*, 189.

[42] These were not entirely independent strains of thought, because MacIntyre was in dialog with the works of Kuhn and Polanyi throughout the development of his views on education in virtue and the nature of tradition. For MacIntyre on Kuhn and, especially, Polanyi, see Alasdair MacIntyre, "Objectivity in Morality and Objectivity in Science," in *Morals, Science, and Sociality*, ed. H. Tristram Englehardt and Daniel Callahan (Hastings-on-Hudson, NY: Hastings Center, 1978), 21–39; Alasdair MacIntyre, "Epistemological Crises and Dramatic Narrative," in *The Tasks of Philosophy: Selected Essays*, vol. 1 (Cambridge: Cambridge University Press, 2006), 3–23; Marjorie Grene, "Response to Alasdair MacIntyre," in *Morals, Science, and Sociality*, ed. H. Tristram Englehardt and Daniel Callahan (Hastings-on-Hudson, NY: Hastings Center, 1978), 40–7; Mark T. Mitchell, "Michael Polanyi, Alasdair MacIntyre, and the Role of Tradition," *Humanitas* 19, no. 1 (2006): 97–125; John Flett, "Alasdair MacIntyre's Tradition-Constituted Enquiry in Polanyian Perspective," *Tradition and Discovery: The Polanyi Society Periodical* 26, no. 2 (2009): 6–20.

[43] Michael Polanyi, *Personal Knowledge: Towards a Post-Critical Philosophy* (Chicago: University of Chicago Press, 1962), 266.

knowledge is a form of belief, belief sustained by trust in a network of other experts and in the authoritative source of one's own education, one's teachers.[44]

Polanyi adds elements to the Aristotelian account outlined above, since his model is undergirded by a recognition of the centrality of tacit knowledge.[45] By tacit knowledge, he means all of the aspects of knowing that cannot be put into words or rules, or at least consciously held while a person performs an act.[46] Knowledge is embodied, but if the actor focuses too much on her body during the performance, she will fail. For example, most people do not really know how they balance on a bike while they ride, and even if they do have this knowledge, it will not help them during the act of riding.[47] When performing at a piano recital, focusing too much on hitting the next note might cause a person to freeze with stage fright because the grasp of the whole performance will be lost.[48]

Much of scientific research consists of embodied, tacit knowledge in much this same way. It is pipetting; it is sterilely moving cells in and out of a fume hood; it is microsurgery on chicken embryos. Far from being an activity of the mind alone, research success depends on having "good hands," since some aspects of research can be done only by people with exceptional dexterity. As in crafts, the tacit aspect of science can be learned only by practice, but practice comes from imitating a master at the craft. A new technique cannot be learned from the Materials and Methods section of a paper, but only by going to a laboratory that performs that technique and undergoing another apprenticeship. The visitor sits next to a graduate student and carefully watches the process. Then the visitor performs the experiment under the eye of the other researcher, before he can say that he actually knows how to do the experiment. Part of this need for imitation and guidance is due to the fact that one cannot write down every instruction, but it is also due to the fact that one does not always know every action that one performs or how to describe how one deploys one's body. This importance of tacit knowledge is one of the reasons that it is so difficult to interpret a failed attempt at replicating another laboratory's experiments. Journal editors and others in the field do not know whether a failed replication results from an unreliable original experiment or because other laboratories do not have the requisite level of skill in performing

[44] Polanyi, *Personal Knowledge*, 53, 163.
[45] Michael Polanyi, *The Tacit Dimension* (Chicago: University of Chicago Press, 2009), 3–25.
[46] Polanyi, *Personal Knowledge*, 49.
[47] Polanyi, *Personal Knowledge*, 49–50.
[48] Polanyi, *Personal Knowledge*, 56.

the experiment. The latter is frequently the first response by a laboratory whose results are challenged by another laboratory. A truly serious attempt at replication requires an exchange of researchers so that outsiders can first learn how to perform an experiment.

Polanyi stretches the concept of embodied knowledge by holding that we incorporate our tools into our body:

> While we rely on a tool or a probe, these are not handled as external objects. We may test the tool for its effectiveness or the probe for its suitability . . ., but the tool and the probe can never lie in the field of these operations; they remain necessarily on our side of it, forming part of ourselves, the operating persons. We pour ourselves out into them and assimilate them as parts of our own existence.[49]

For Polanyi, a conceptual apparatus is another tool one uses for scientific research; background conceptual beliefs, tacit intellectual knowledge, become embodied. Science can be learned only through an apprenticeship that transforms the apprentice by mimetic appropriation of a master's tacit knowledge: "An art which cannot be specified in detail cannot be transmitted by prescription, since no prescription for it exists. It can be passed on only by example from master to apprentice." Therefore, one must accept authority. "You follow your master because you trust his manner of doing things even when you cannot analyse and account in detail for its effectiveness."[50] This submission does not crush progress, though. "Every thoughtful submission to authority is qualified by some however slight opposition to it." Traditions are "our own interpretations of the past, at which we have arrived within the context of our own immediate problems."[51] This appropriation allows one to transform the community's tacit knowledge through further research.

Kuhn similarly focuses on the importance of scientific education through his concept of the scientific paradigm. Scientific communities exist under the influence of a particularly successful theoretical achievement, such as Newton's *Principia*.[52] Such a paradigm provides a model of how to go about the process of solving scientific puzzles. It is a disciplinary

[49] Polanyi, *Personal Knowledge*, 59.
[50] Polanyi, *Personal Knowledge*, 53.
[51] Polanyi, *Personal Knowledge*, 160. MacIntyre described the authoritarian overtones of Polanyi's tradition as Burkean in his early work, such as MacIntyre, "Objectivity in Morality and Objectivity in Science." As he grew to emphasize the role of authority, MacIntyre became less critical of Polanyi.
[52] Thomas Kuhn, *The Structure of Scientific Revolutions*, 3rd edition (Chicago: University of Chicago Press, 1996), 10.

matrix of both mental aspects, such as concepts and metaphors, and more operative aspects, such as acceptable techniques and equipment.[53] It marks out the domain of worthwhile problems, which are solved as the paradigm's original problems were solved. The foundational work acts as an exemplar for later work.

The textbook educates the student into a paradigm. Kuhn explains, "Science students accept theories on the authority of teacher and text … The applications given in texts are not there as evidence but because learning them is part of learning the paradigm at the base of current practice."[54] The student grows in knowledge and ability not by learning the conceptual resources provided by a paradigm but by the practice of problem solving.[55] Examples of paradigmatic problems allow the student to work through the issues that lie at the core of the paradigm and to apply the paradigm to new problems. For genetics, the beginning student works through the results of Mendel's experiments on peas, even though they have little direct relevance to current research and are rather abstract. These problem sets give the student an example to build from as she confronts more difficult issues. Thus, Kuhn's education is a submission to authority in order to gain the conceptual resources of the community.[56]

Philosophy of science influenced MacIntyre's discussions of the rationality of traditions,[57] but the education of scientists relates to MacIntyre's concept of moral formation as well. MacIntyre explicitly argues that the power and success of modern natural science is due to its retention of a structure of authority.[58] Whereas the humanities have seen the blossoming of any number schools of thought, the natural sciences have always been careful to exclude anyone who drifts too far away from standard methods and paradigms, unless the standard paradigm degenerates into a recognizable crisis. Even then, Polanyi and Kuhn see battles between scientific theories as much about gaining control of institutions as actually convincing others of one's theories.[59] In this way, natural science retains something akin to the religious test for teaching that MacIntyre regrets that the modern liberal university has lost.[60] Such tests ensure a dedication to the unity of knowledge under a framework guaranteed by authority.

[53] Kuhn, *The Structure of Scientific Revolutions*, 181–204.
[54] Kuhn, *The Structure of Scientific Revolutions*, 80.
[55] Kuhn, *The Structure of Scientific Revolutions*, 46–7.
[56] For authoritarian dangers in Kuhn's conception of science, see Steve Fuller, *Thomas Kuhn: A Philosophical History for Our Times* (Chicago: University of Chicago Press, 2000).
[57] MacIntyre, "Epistemological Crises and Dramatic Narrative."
[58] MacIntyre, *Whose Justice? Which Rationality?*, 400.
[59] Kuhn, *The Structure of Scientific Revolutions*, 152; Polanyi, *Personal Knowledge*, 152.
[60] MacIntyre, *Whose Justice? Which Rationality?*, 399.

Thus, many aspects of scientific training reflect MacIntyre's Aristotelian model of moral formation through craft.

Reductionism in Scientific Practice

While the previous section suggests ways in which scientific training may push against the problems of entrepreneurialism in science through a craft-like process of formation in virtues, there are other aspects of the practice of science that might accentuate the issues presented in Chapter 2. Polanyi's suggestion that the researcher's conceptual tools serve as an extension of the self points to where the principle dangers might lie. If conceptual schema become embodied as tacit knowledge, then scientific training will in fact undermine a training in the virtues if its conceptual resources conflict with the necessary teleological understanding of the world that undergirds Aristotelian virtue. As this section will argue, that is exactly the case with the concepts of contemporary science, which encourage a reductionist understanding of the world. Indeed, it was the development of strands of natural science that undermined a teleological model of nature, not merely the discussions of moral philosophers. Through the daily practices of science, this reductionist, anti-teleological worldview becomes embodied in the character of the researcher, predisposing her to accept the commodification of the world. The mathematical tools used by science and the specific experimental methods on which science draws shape modern subjects in a way inhospitable to a teleological view of the world, allowing them to see objects in the world as infinitely manipulable. This analysis, if correct, has serious implications for the proposed Aristotelian solution. It raises the question of how to deal with a situation where the habits necessary for achieving the internal goods of a *techne* seem to conflict with the virtues necessary for a full unity of life.

Technological Reductionism in the Scientific Worldview

As we have seen, *techne*, the communal craft tradition of making and knowing that includes contemporary science, is opposed to technique, utilitarian methods of measurable effectiveness. However, these *technai* have changed through their alliance and cooperation with science, especially mathematical physics, to become technology.[61] The concern that numerous authors raise is that the intellectual habits formed by the

[61] George Grant, "Knowing and Making," in *The George Grant Reader*, ed. William Christian and Sheila Grant (Toronto: University of Toronto Press, 1998), 411.

mathematical bent of modern science are ultimately opposed to the teleological vision of the world the Aristotelian model holds as necessary for a coherent character.[62] The concern is that habits and thought styles of natural science do not form an Aristotelian subject.

Martin Heidegger presents what is perhaps the most influential form of this critique.[63] Like Aristotle, Heidegger agrees that at first sight technology, meaning technical artifacts and methods, is morally neutral.[64] One must look beyond its surface appearance, however, to get a true insight into technology's essence.[65] As Weber argued, behind all the technological manifestations in society lies a view of the world as a mathematically calculable array of objects and forces to be manipulated: "Modern science's way of representing pursues and entraps nature as a calculable coherence of forces."[66] Moderns see natural objects not as having their own ends but as merely a standing reserve of energy to be ordered for human ends, making the entire world a network for use.[67] Beyond the threat to other organisms and the environment presented by this development, the danger that Heidegger most fears is that humans will begin to conceptualize themselves as merely such means to be manipulated through mathematical technology, losing their vocation to respond to the call of being and to "experience the call of a more primal truth."[68] Heidegger criticizes the rise of Weberian managerial technique, fearing that the triumph of instrumental reason will obscure humanity's vocation for truth.

Natural science, especially in the form of mechanistic physics, played an essential role in defeating teleology.[69] Yet these scientific developments

62 On the need for MacIntyre to address the role of modern science in his analysis, see Thomas Hibbs, "MacIntyre's Postmodern Thomism: Reflections on Three Rival Versions of Moral Enquiry," *The Thomist* 57, no. 2 (1993): 277; Terry Pinkard, "MacIntyre's Critique of Modernity," in *Alasdair MacIntyre*, ed. Mark Murphy (New York: Cambridge University Press, 2003), 176–200.

63 Here I will only deal with the form of Heidegger's critique of technology in "The Question Concerning Technology" and "The Age of the World Picture," since it is the form of his argument that provides the most grounds for discussion with MacIntyre and Foucault. See Martin Heidegger, "The Question Concerning Technology," in *The Question Concerning Technology, and Other Essays* (New York: Harper Torchbooks, 1977), 3–35; Martin Heidegger, "The Age of the World Picture," in *The Question Concerning Technology, and Other Essays* (New York: Harper Torchbooks, 1977), 115–54. For another analysis of his later thought on technology from a theological perspective, see Brian Brock, *Christian Ethics in a Technological Age* (Grand Rapids, MI: William Eerdmans, 2010).

64 Heidegger, "The Question Concerning Technology," 5.

65 Heidegger, "The Question Concerning Technology," 6.

66 Heidegger, "The Question Concerning Technology," 21.

67 Heidegger, "The Question Concerning Technology," 16–17.

68 Heidegger, "The Question Concerning Technology," 28.

69 This narrative is discussed by many authors. See, for example, John Hedley Brooke, *Science and Religion: Some Historical Perspectives* (New York: Cambridge University Press, 1991), 158–205.

have deeper roots in late medieval theological debates. Here I will follow the broad outline of this historical narrative provided by Hans Blumenberg in *The Legitimacy of the Modern Age*. There are other variants of this narrative and criticisms of it, but he provides a version that is sympathetic to modern developments against teleology, and thus at least is somewhat even-handed in his approach.[70] In this genealogy of technology, the rejection of Aquinas's marriage of Augustinian Christianity and Aristotelianism that began with the Condemnations of 1277 led to a divorce between theology and philosophy.[71] A greater emphasis on divine omnipotence and fallen human nature by voluntarist theologians led to doubt in our natural capacities to gain knowledge of the world, especially of the nature and ends of creatures and processes. "The disappearance of teleological protections that had been part of the concept of nature means that man has to adjust himself to coming to terms with a nature that is not adjusted for his benefit."[72] If God could make an infinite number of worlds that would be good, then the things and processes of this existing world cannot be explained by the principle of sufficient reason. God could have made things otherwise, so perhaps the object in front of us does not have any necessary purpose. Moreover, even if things had purposes, then God would not necessarily make knowledge of these purposes accessible to humans.[73]

In response to this growing doubt in human capacities to know the world, natural philosophers sought for a way to reason about such an uncertain nature. They found it in the method of hypothesis.[74] Astronomy had long used the method of hypothesis, and philosophy adopted tools of hypothesis to describe the rest of nature, tools that were essentially mathematical.[75] Ancient astronomers never could gain access to their objects of study given their residence in the sublunary realm, but merely attempted "to render the unknown and inaccessible mechanism of the goings on in the heavens sufficiently calculable to meet human needs."[76] Yet at an early

[70] E.g., Servais Pinckaers, *The Sources of Christian Ethics*, 3rd edition, trans. Mary Thomas Noble (Washington, DC: Catholic University of America Press, 1995); MacIntyre, *Three Rival Versions of Moral Enquiry*, 149–69. Blumenberg's narrative is also discussed in Northcott, *The Environment and Christian Ethics*, 57–59.

[71] Hans Blumenberg, *The Legitimacy of the Modern Age* (Cambridge, MA: MIT Press, 1985), 159–79.

[72] Blumenberg, *The Legitimacy of the Modern Age*, 199.

[73] Blumenberg, *The Legitimacy of the Modern Age*, 188–90.

[74] Blumenberg, *The Legitimacy of the Modern Age*, 199.

[75] Blumenberg, *The Legitimacy of the Modern Age*, 200. For a more thorough discussion, see Pierre Duhem, *To Save the Phenomena: An Essay on the Idea of Physical Theory from Plato to Galileo* (Chicago: University of Chicago Press, 1969).

[76] Blumenberg, *The Legitimacy of the Modern Age*, 200.

stage, it became evident that multiple mathematical models could account for the deviations in the movements of the stars, so the choice between models would have to be made on metaphysical rather than astronomical grounds. A conventionalism arose in astronomy, holding that a theory could be predictive but not necessarily true, and thus decisions between models must be made on probable grounds. This conventionalism was affirmed by the Aristotelian tradition, including Aquinas, and it was the ground on which Cardinal Bellarmine urged Galileo not to claim the truth of his theory.[77] Galileo claimed truth and, along with Cartesian metaphysics, introduced a mechanistic model of the world antithetical to teleology.

The Copernican revolution raises problems for the teleological ethical project.[78] First, critics of modernity want the unification of knowledge rather than disciplinary fragmentation. However, they fail to solve the question of how to unite other disciplines to mathematical science in a way that will not lead to reductionist tendencies and the rejection of teleology.[79] Second, mathematical science trains one in habits of mind that reject teleology and view the world instrumentally. It is an example of a *techne* in which the apprenticeship will not lead to the form of intellectual character that MacIntyre wants.

To address the first problems, three solutions to the difficulties of unifying mechanistic physics with teleological systems seem to be available, as MacIntyre's engagement with scientific knowledge suggests. Stephen Pope charts the gradual development of MacIntyre's position.[80] In *After Virtue*, he rejects Aristotle's metaphysical biology, instead grounding teleology in human practices and communities, which solves the problem by separating the human from the natural sciences.[81] This move fails because by rejecting a foundation of ethics in human animality, it moves far from

77 Duhem, *To Save the Phenomena*.

78 While MacIntyre acknowledges that the rise of mechanistic physics was intertwined with the rejection of Aquinas and teleology, he does not fully address the problems this raises for his project. Generally, he affirms these changes, repeatedly using the Copernican revolution as the example of how the rationality of traditions progresses (e.g., MacIntyre, "Epistemological Crises and Dramatic Narrative"; MacIntyre, *Whose Justice? Which Rationality?*, 365). For MacIntyre, despite being a defeat for the Aristotelian method, Galileo's success was a victory for the Aristotelian narrative of inquiry. See Alasdair MacIntyre, "First Principles, Final Ends, and Contemporary Philosophical Issues," in *The Tasks of Philosophy: Selected Essays*, vol. 1 (New York: Cambridge University Press, 2006), 163.

79 I am not arguing about the best way to relate science to theology, or science to philosophy, but merely discussing the intellectual habits formed by certain ways of doing science.

80 Stephen J. Pope, *Human Evolution and Christian Ethics* (New York: Cambridge University Press, 2007), 271–3.

81 MacIntyre, *After Virtue*, 196.

the teachings of Aristotle and Aquinas.[82] Loss of a biological basis for ethics in human nature destroys the unity of knowledge. This recognition still does not solve the problem of combining reductionism with teleology.

A second possible response is that each practice progresses to truth or excellence at its own rate through its own crises. Against Kuhn, who, in arguing that science has no tendency to progress to a true account of nature, suggests that relativistic physics is in many ways closer to Aristotle than to Newton, MacIntyre responds by saying that Newtonian physics was a necessary step on the approach to truth.[83] The problems that were solved by Einstein could not have arisen in Aristotelian physics. Thus, seemingly nonteleological sciences can be seen as necessary points of passage to a better teleological account of truth, which helps at the level of historical narrative but not in terms of how to address our current problems in shaping an intellectually coherent mindset out of a mechanistic science combined with a teleological ethics.

A third maneuver that MacIntyre seems to make is to argue for a hierarchical arrangement of nature. Aristotle's

> is a universe structured in a hierarchical way ... and each level of the hierarchy provides the matter in and through which the forms of the next higher level actualize and perfect themselves. The physical provides the material for biological formation, the biological the material for human formation.[84]

Higher levels are not reducible to the lower levels of being. This conception of the universe was not necessarily defeated when medieval mechanics was replaced by the work of Galileo and Newton, since one can make use of knowledge gained from the lower level in making higher-level arguments without subordinating the higher level to it. One can use dolphin sociality to argue that our social nature is based in our animal nature without denying the differences that exist between human and animal sociality. This move is similar to broader arguments for emergence as an explanatory principle in science.

Such arguments better confront the problem of combining mathematical science and teleological ethics, but three difficulties remain. First, as many philosophers have noted, there are significant conceptual issues with

[82] MacIntyre recognizes this problem in Alasdair MacIntyre, *Dependent Rational Animals: Why Human Beings Need the Virtues* (Chicago: Open Court, 1999), x. Thus, he begins to make use of biological research on animal, especially dolphin, behavior in his work, but without resolving the theoretical difficulties involved (21–51).

[83] MacIntyre, "Epistemological Crises and Dramatic Narrative," 21.

[84] MacIntyre, *Whose Justice? Which Rationality?*, 101.

the idea of emergence itself. It is not entirely clear what it means or what it explains. As Robert Spaemann argues, emergent properties mean

> properties whose coming-to-be regularly arises out of certain material combinations but cannot be understood as combinations of the properties of the uncombined materials. But properties have to be properties of something. The coming-to-be of a new something, individuation cannot be described as an emergent property.[85]

Further, it is not clear how one would engage it in the practice of science, because it is not used by many scientists themselves.[86] Second, if one looks to the supposedly higher levels of organization, one finds that research on them is becoming increasingly mathematical and reductionist. Biology is a prime example. Finally, such solutions do not deal with the problem of the habits of mind induced through the use of mathematical technique at all levels of organization and by the necessary education in physical science to understand biological science. I will address these last two concerns in order.

Up to this point in the discussion, the main dangers to teleology seemed to lie in mathematical physics, but biomedical science also gives rise to the mathematical, instrumentalist, and fragmented habits of mind. At first glance, biology seems to be more open to teleology than physics and chemistry. Even Heidegger thought that biology might be held exempt from the expansion of technological thought since the study of organisms requires it to recognize inherently purposive activity.[87] This recognition was the basis of Ernst Mayr's defense of a teleonomic heuristic in biology.[88] To understand a bird's migration, it is not enough to unveil the biological, physical, and chemical mechanisms that lead a bird to fly a certain route at a certain time of year, since one must also understand the migration as purposeful, in that it is a search for new feeding grounds. Mayr refuses to call this purpose teleological, since this term is closely tied to ideas of a cosmic teleology.[89] Instead, he calls it teleonomic: a goal-directed behavior that arises from a program, a metaphor derived from

85 Robert Spaemann, *Persons: The Difference between "Someone" and "Something,"* trans. Oliver O'Donovan (New York: Oxford University Press, 2017), 43.

86 Thomas Nagel, *Mind and Cosmos: Why the Materialist Neo-Darwinian Conception of Nature Is Almost Certainly False* (New York: Oxford University Press, 2012).

87 Heidegger, "The Age of the World Picture," 120.

88 Ernst Mayr, "The Multiple Meanings of Teleological," in *Toward a New Philosophy of Biology: Observations of an Evolutionist* (Cambridge, MA: Harvard University Press, 1989), 54.

89 Mayr, "The Multiple Meanings of Teleological," 40.

computer programs.[90] He insists on the importance of the concept of the program, looking to the program embedded in an organism's DNA and the programs embedded in our central nervous system acquired from learning. To defend this notion, he draws on Max Delbruck's claim that DNA is the equivalent of Aristotle's form.[91] However, this embrace of form and teleonomy is not really a hearkening back to Aristotle, but is rather the embrace of information technology and cybernetics, which allows Mayr not to differentiate between organisms and machines like computers.[92] Many commentators see biology as a barrier to the expansion of mechanistic thought, but as the equation between organism and computer indicates, even an embrace of teleology gives openings for an instrumental mindset.

Similarly, Polanyi prefers to see the purposes evident in organisms as operational principles, normative "rules of rightness" directed toward purposes, as one would have operational principles in a machine. "The class of things defined by a common operational principle cannot be even approximately specified in terms of physics and chemistry."[93] It is an explanatory level beyond the particles of physics since the parts of the organism act as a mechanism to achieve an end. There is an overlap between the organism and machine insofar as the organism is understood as such a mechanism. "The body of an animal also functions to a certain extent as a machine."[94] Polanyi holds that biology and medicine are the closest sciences to technology, even overlapping with it, since they both can be understood in terms of engineered design to achieve ends.[95] The important conclusion that results from these arguments is that since biology, like technology, deals with operational principles, biological processes should be open to patent: "Claims for hundreds of patents could be found by describing the rational interaction of the animal's organs in the service of its various interests." This is an example of how this reductionist

[90] Mayr, "The Multiple Meanings of Teleological," 45.

[91] Mayr, "The Multiple Meanings of Teleological," 49, 62. This line of reasoning is cautiously drawn on by International Theological Commission, "Communion and Stewardship: Human Persons Created in the Image of God," 2004, para. 30, www.vatican.va/roman_curia/congregations/cfaith/cti_documents/rc_con_cfaith_doc_20040723_communion-stewardship_en.html.

[92] Mayr, "The Multiple Meanings of Teleological," 49. See also Lily Kay, *Who Wrote the Book of Life?: A History of the Genetic Code* (Stanford, CA: Stanford University Press, 2000), 38.

[93] Polanyi, *Personal Knowledge*, 321.

[94] Polanyi, *Personal Knowledge*, 332.

[95] Polanyi, *Personal Knowledge*, 342.

rationality makes the entrepreneurial scientist and his patents possible.[96] Boundaries between mechanistic and teleological thought blur.

Molecular biology has been heavily influenced by mathematical thought since its foundation.[97] The most innovative subdisciplines grew by the movement of physicists into biology during the middle of the last century. These new entrants transformed biological thought, in part by introducing information theory into genetics, a heuristic that was tremendously important for unraveling how genes function in the cell, existing as DNA that is transcribed into RNA that is translated into proteins that then effect functions in the cell. Information theory calls on fundamentally mathematical ideals, even if used in a largely metaphorical manner in biology.

Current research indicates that this picture of cell function is too simplistic.[98] The genes in DNA do not simply rule the cell, directing its action through its stored information. Instead, cellular processes feed back onto the DNA, regulating it. DNA is transformed by the cell's history through epigenetics, and much of the cell's activity is structured by the interactions of proteins. The cell is now viewed as a system of interacting parts in a more holistic synthesis. However, this more holistic vision has merely reinforced the mathematical tendencies of biology. This complexity has driven the collection of vast amounts of data on gene sequence, gene regulation, and protein interaction, big data that are mined by an army of programmers. The view of the cell as a system has brought a new influx of physicists and mathematicians who are trained to model complex systems. Models of signaling pathways in the cell now look like diagrams of extraordinarily complex circuits. Many biologists approach the study of life as fundamentally a matter of calculation, a worldview prepared by training in math, physics, and chemistry.

Faced with this mass of uninterpreted data, the most enthusiastic promoters of the digital revolution argue that we should do away with the human element altogether.[99] If humans cannot process this much data, why not leave the job to computers? Science can do without hypotheses altogether, since, with enough data, programs should be able to identify patterns and detect correlations. Science can yield completely to technology.

[96] Polanyi, *Personal Knowledge*, 359. Interestingly enough, however, both Polanyi and Hayek, though neoliberal in orientation, feared the monopolistic effects of privately held patents. Both argued for publicly held patents with some basic royalties granted from low licensing fees. See Michael Polanyi, "Patent Reform," *Review of Economic Studies* 11, no. 2 (1944): 61–76; Friedrich A. von Hayek, *Individualism and Economic Order* (Chicago: University of Chicago Press, 1948), 114.

[97] Kay, *Who Wrote the Book of Life?*.

[98] For discussion, see Lenny Moss, *What Genes Can't Do* (Cambridge, MA: MIT Press, 2004).

[99] Chris Anderson, "The End of Theory: The Data Deluge Makes the Scientific Method Obsolete," Wired, June 23, 2008, www.wired.com/2008/06/pb-theory/.

Most actual scientists, even those researching bioinformatics, whose biological investigations are grounded in quantitative theory and method, reject such an extreme vision of autonomous Big Data, insisting that research will always require human creativity and insight. The most fervent rejection has been expressed, surprisingly, by the most reductionist of molecular biologists. Craig Venter, who launched a private-sector alternative to the government-sponsored Human Genome Project, has criticized the focus on emergent properties arising from "vast numbers of interacting chemical processes forming interconnected feedback cycles," calling it "a new kind of vitalism," the belief that life must be explained by a nonmaterial force.[100]

Sydney Brenner, one of the founders of molecular biology and a Nobel Prize winner for his work on roundworms as model systems, spent much of the first decade of this century criticizing "omics" sciences in lectures and papers. As Brenner argued in a 2009 interview with the journal *Studies in History and Philosophy of Science*, the current trend in molecular biology, "systems biology," "is technology driven ... [and] does not have to pose any hypothesis. In other words, it claims to release people from thinking. You do not have to think, you just make an array and get a lot of numbers."[101] Today's research paradigm brings to mind the doubts about midcentury physics, particularly quantum mechanics, expressed by thinkers such as Martin Heidegger and Hannah Arendt: they were concerned that experimental data no longer appeared in the form of representational theories. Instead, the results of experiments were merely "orderable as a system of information ... It seems as though causality is shrinking into a reporting."[102] There is no possibility of a teleological worldview if there is no interest in causality in contemporary conceptions of biological or physical understanding.

Laboratory Practices and Reductionism

A common response to these problems of conceptual reductionism is to argue that one can separate methodological reductionism from a fuller conceptual embrace of reductionism. Stephen Pope provides a taxonomy

100 J. Craig Venter, *Life at the Speed of Light: From the Double Helix to the Dawn of Digital Life* (New York: Penguin Books, 2013), 17.

101 Soraya de Chadarevian, "Interview with Sydney Brenner," *Studies in History and Philosophy of Science Part C: Studies in History and Philosophy of Biological and Biomedical Sciences* 40, no. 1 (March 2009): 65–71.

102 Heidegger, "The Question Concerning Technology," 23; Hannah Arendt, *The Human Condition* (Chicago: University of Chicago Press, 1998), 285–90.

of different forms of reductionism in *Human Evolution and Christian Ethics*.[103] There is ontological reductionism that addresses metaphysics, indicating that higher levels of organization in nature "are nothing more than a particular way in which simpler traits or entities are organized."[104] Epistemological reductionism is similar, attempting to explain higher levels of organization through the interactions of lower levels of organization. Finally, there is methodological reductionism, which is simply "a research strategy for examining wholes in terms of their constituent parts."[105] When used well, researchers employing methodological reductionism recognize the limited nature of its claims and do not try to use it to overthrow other ways of thinking about the world. Pope, with many other theologians, argues that "Christian ethics can accept the methodological reductionism of evolutionary theory while rejecting its epistemological and ontological reductionism."[106] There is no logical necessity that links methodological with other forms of reductionism.

While I agree with Pope that there is no *logical* entailment between methodological and ontological reductionism, there are problems when one takes seriously the way that everyday practices shape one's character. If one accepts Polanyi's notion that conceptual schema become embodied, tacit knowledge, then daily engagement with reductionist practices, even if one tries to hold them only in the form of methodological reductionism, will have the effect of shaping one in a more generally reductionist worldview. Daily laboratory practices reinforce the mental habit of seeing the world as infinitely operable instead of teleological. These problems have been most thoroughly investigated in the practice of medicine. In his book *The Anticipatory Corpse*, Jeffrey Bishop builds on Foucault's work in *The Birth of the Clinic* to argue, with many others, that contemporary problems in medicine result from making the dead body the paradigmatic body.[107] Medicine, following physiology, has replaced the final causes of the living body with the material and efficient causes of the particles of the corpse.[108] These conceptual issues have translated into daily practices of training and treatment: dissections in medical school, the clinical gaze that ignores the patient's words in favor of test results, or the shape of the ICU that treats

103 Pope, *Human Evolution and Christian Ethics*, 56–75.
104 Pope, *Human Evolution and Christian Ethics*, 69–70.
105 Pope, *Human Evolution and Christian Ethics*, 57.
106 Pope, *Human Evolution and Christian Ethics*, 72.
107 Jeffrey Bishop, *The Anticipatory Corpse: Medicine, Power, and the Care of the Dying* (Notre Dame, IN: University of Notre Dame Press, 2011), 21.
108 Bishop, *The Anticipatory Corpse*, 20.

the body as merely mechanistic functions rather than in terms of personal presence. Concepts are translated into practices that then form the person.

This argument applies to biomedical research as well. In the laboratory, the researcher takes the dead body of his model organism, dissects it, and then fragments it into its various components. Over my eight years spent researching the genetics of vertebrate embryonic development, an average day in the laboratory might consist of selecting one of my model organisms, a pregnant mouse, for example, humanely killing it, tearing open its stomach to dissect out the developing embryos in its uterus, and then, over the next few days, dissociating these embryos to access information about the gene or protein of interest. By dissociating, I mean that I might dissolve the limb of the embryo, extract proteins, run them out on a gel, and probe it with an antibody, ending with a clear line on a piece of film. Or I might embed the embryo in paraffin, slice it into 6-micron-thick sections, and stain these sections to pinpoint where a gene turns on in a tissue. Even though the researcher might never consciously embrace reductionism, these daily embodied practices of laboratory life, along with the theoretical frameworks provided in classrooms and journal articles, train him in habits of perceiving organisms as fundamentally networks of genes and proteins that can be dissociated, altered, and reengineered. Method forms a way of being in and interacting with the world.[109]

Of course, knowledge gained through fragmentation must be tested in the whole organism. Yet one can test the function of a component of the organism only by changing the organism. As Paul Rabinow writes in regard to the Human Genome Project:

> The object to be known – the human genome – will be known in such a way that it can be changed. This dimension is thoroughly modern; one could even say that it instantiates the definition of modern rationality. Representing and intervening, knowledge and power, understanding and reform, are built in, from the start, as simultaneous goals and means.[110]

The tools of knowledge are the tools of technological transformation. For example, one tests gene function by creating transgenic mice that have a mutated form of the gene or express the gene in the wrong place or at the wrong time. Knowledge through operability is brought to its highest point

[109] Many embodied laboratory practices do support claims of *technai* forming virtue, although these are not the focus of my discussion. The extreme difficulty of microsurgery or the monotony of hours of transferring liquids from tube to tube builds patience. Experiments that last through the night build temperance.

[110] Paul Rabinow, "Artificiality and Enlightenment: From Sociobiology to Biosociality," in *Essays on the Anthropology of Reason* (Princeton, NJ: Princeton University Press, 1996), 93.

in synthetic biology, which dreams of creating an organism from scratch with purposes designed by us.[111] In this goal, synthetic biology echoes the aims of the pioneering developmental biologist Jacques Loeb, who discovered artificial parthenogenesis, the technique for inducing an unfertilized egg to begin the process of development, in order to further a "technology of living substance."[112] Biology reaches the heights of technique at this point.

Escaping the vision of dead matter does not necessarily get us back to natural teleology, as analyses of tissue culture technologies show.[113] Tissue culture was originally envisioned as a way to work experimentally with live organisms.[114] The Nobel Prize–winning scientist Alexander Carrel did not merely want to see sectioned dead tissue, but to experiment on vital tissue, so he developed the means to work with live tissue outside the body to develop a technological form of life.[115] This manipulation of living tissue required its isolation from the body, reinforcing the vision of the body as open to fragmentation. Speaking about one cell line, a researcher says, "Scientists don't like to think of HeLa cells as being little bits of Henrietta because it's much easier to do science when you dissociate your materials from the people they come from."[116]

Regenerative medicine and its stem cell technologies emphasize this fragmentation and operability to a greater degree than other tissue culture technologies. For induced pluripotent stem cells (iPSCs), the goal is to take cells from an adult, and then use genetic or chemical techniques to force them into a pluripotent state, the state of the cells of the early embryo.[117] Pluripotent cells have the capacity to form any cell of the adult human body, and thus could provide the raw material for treatment of diseases like diabetes or Parkinson's. iPSCs are transformed into this early stage, an

111 Paul Rabinow and Gaymon Bennett, *Designing Human Practices: An Experiment with Synthetic Biology* (Chicago: University of Chicago Press, 2012).

112 Philip Pauly, *Controlling Life: Jacques Loeb and the Engineering Ideal in Biology* (New York: Oxford University Press, 1987), 4.

113 Here I draw on the insights of Paul Rabinow, "Severing the Ties: Fragmentation and Dignity in Late Modernity," in *Essays on the Anthropology of Reason*, 129–52; Cathy Waldby and Robert Mitchell, *Tissue Economies: Blood, Organs, and Cell Lines in Late Capitalism* (Durham, NC: Duke University Press, 2006); Hannah Landecker, *Culturing Life: How Cells Became Technologies* (Cambridge, MA: Harvard University Press, 2007); Melinda Cooper, *Life as Surplus: Biotechnology and Capitalism in the Neoliberal Era* (Seattle: University of Washington Press, 2008).

114 Landecker, *Culturing Life*, 32.

115 Landecker, *Culturing Life*, 3.

116 Rebecca Skloot, *The Immortal Life of Henrietta Lacks* (New York: Crown Publishers, 2010), 216.

117 Junying Yu et al., "Induced Pluripotent Stem Cell Lines Derived from Human Somatic Cells," *Science* 318, no. 5858 (2007): 1917–20.

undifferentiated mass of cellular material that will hopefully differentiate into the needed cell types. Biological material is fragmented from the human body and changed into a standing reserve of biological potentiality to be drawn on as needed.

The conceptual and practical work of regenerative medicine encourages what the Heideggerian critique would call a technological world picture and would work against teleological conceptualizations of the world. It encourages habits of action that engage the world as manipulable and fragmentary. The scientist entrepreneur can more easily commodify creatures in such a manipulable and fragmented world. While MacIntyre does not claim that all habits that result from a practice are virtues,[118] it is difficult to see how many of the habits of mind and action formed by contemporary biotechnology fit into his schema of moral formation. They seem to play a greater role than mere skills in his system, yet they would not be true virtues on his account. Yet he accepts biology as a true practice and *techne*, whose results must be accepted as part of a unified field of knowledge. Further development of his thought on science as a practice is necessary to solve this dilemma.

Distinguishing Entrepreneurial from Older Modes of Reductionism

The argument of this chapter has been that the necessary precondition of entrepreneurial science is a reductionist vision of the world, a worldview shaped by the daily habits of scientific practice. However, because this view of the world was established early in the development of modern science, it does not entail a neoliberal framework for structuring the world. By this, I mean that viewing and dealing with the body as fragmentable and as a mathematically calculable and manipulable network of materials and forces does not lead automatically to viewing biological materials as patentable and to the scientist entrepreneur. Nor does an economic setting by itself necessarily lead to a certain kind of science. However, viewing the body as a collection of fragments and information does allow for a commodification of those distinct fragments and thus for the scientist entrepreneur. There is no necessary relationship, and thus no economic or technological determinism, but reductionism does serve as a condition of the possibility of entrepreneurial science.

[118] MacIntyre, *After Virtue*, 275; helpfully noted by Porter, "Tradition in the Recent Work of Alasdair MacIntyre," 41.

The distinction between two cases of medical researchers removing cancerous tissue from patients and transforming them into cell lines occurring before and after the entrepreneurial transition in science shows that one can engage this reductionist vision of the body under two very different structures of scientific practice. Both involve removing a part of a patient's body, but only the latter involved the transformation of that tissue into a patented resource. Both raise moral concerns and questions of the character of the scientist, but of different sorts.

The first case is the well-known one of HeLa cells.[119] In 1951, Henrietta Lacks, a young black woman, was biopsied and diagnosed with cervical cancer. She was treated at Johns Hopkins Hospital, a hospital established to provide treatment for those who could not afford it, including a large portion of Baltimore's black population. It also was committed to medical research, and part of her biopsy was given to George Gey, who had tried to establish an immortal human cell line for many years to aid in finding a cure for cancer and other diseases. Human cells generally last for only a limited number of cell divisions in culture, so Gey searched for cells that could grow in culture indefinitely. While Henrietta Lacks died eight months after her biopsy, the cells of her biopsy gave rise to an immortalized line, HeLa, which could grow indefinitely in culture. After an initial characterization, Gey distributed these cells to many researchers, and they became critical to the development and testing of the polio vaccine. Since that time, they have become a standard instrument of cell biology.

The history of these cells raises two ethical issues related to my argument.[120] In both research and public rhetoric, the cells are forever tied to the woman from whom they were taken. "The information gleaned from

[119] I draw my account from Landecker, *Culturing Life*, 107–79; Skloot, *The Immortal Life of Henrietta Lacks*.

[120] I pass over two equally important ethical concerns that are not central to my argument in this chapter. First, the researchers did not receive truly informed consent for the use of the biopsy in research, for the autopsy performed after her death, nor for the later investigation of the Lacks family genetics. While we must see this lack of informed consent as deplorable today, the researchers acted according to the standards of the time. This issue is well dealt with by standard bioethics and has recently come to a negotiated conclusion through an agreement between the NIH and the Lacks family, so I will not pursue it here. Second, there are problems of racial injustice, from the fact that she was treated at Johns Hopkins Hospital in the first place as a result of structural economic injustices and racial discrimination at other hospitals, through to portrayals of Lacks and her cell line in light of negative stereotypes during a later contamination problem, on to the continuing poverty of her descendants. For discussion of both issues, see Skloot, *The Immortal Life of Henrietta Lacks*.

cells is useless unless it eventually relates back to the biology and then the pathology of the patient."[121] The cells are tied to her, her family, and her history. On the other hand, the cells are treated as mere research material. They have been sent into space, cloned, combined with mouse cells, and exposed to atomic explosions. This combination of personal connection and objectification troubles Henrietta Lacks's daughter, who has dreams of her mother cloned or combined with mice. It exemplifies the distanciation that scientists must make between their materials and the actual whole person or animal.[122]

The second issue concerns economic justice. Beginning in the 1980s, the Lacks family began asking whether they deserved a share in the profits made from research on HeLa cells.[123] In contrast to what would have occurred today neither Gey nor Johns Hopkins made money from the 1951 isolation of HeLa cells. He did not patent the cell line, nor did he hold it back to perform research on it himself, but rapidly distributed it to colleagues. It was a very different ethics of research. The mass production of HeLa cells occurred through a public program located at the Tuskegee Institute, but was later transferred to private companies for greater convenience. While the first program was nonprofit, the companies did make money from the cells, but it was money from production and sale of a commodity rather than a patented item. A researcher could always get a sample of HeLa cells from another laboratory that already had them. Gey believed he was acting for knowledge and the public good, and thus the profit motive was not involved.

Such concerns were central to a second case. In 1976, as the intellectual property and financial reforms that introduced the scientific entrepreneur were being put in place, David Golde removed John Moore's spleen to treat his hairy cell leukemia.[124] Unbeknownst to Moore, Golde established a cell line from Moore's T lymphocytes: the Mo line. Following the newly emerging entrepreneurial ideal, Golde secured a patent on the by-products of the cell line and became a consultant of the company to which the

[121] Landecker, *Culturing Life*, 175.

[122] For a discussion of the problems raised by this dissociation, see Paul Scherz, "Fragments of the Body in Christian, Bioethical and Social Imaginaries," *Studies in Christian Ethics* 30, no. 4 (November 1, 2017): 449–63.

[123] Landecker, *Culturing Life*, 172.

[124] My account draws on Rabinow, "Severing the Ties: Fragmentation and Dignity in Late Modernity"; Waldby and Mitchell, *Tissue Economies*, 88–109.

patent was licensed. When Moore discovered these events many years later when Golde was asking him to sign consent forms that were more explicitly geared toward Moore approving the commercial exploitation of his cells, Moore sued. He thought it unjust for Golde to make so much money from his tissues, and he wanted a share of Golde's profits. The California Supreme Court ruled against Moore, arguing that the cells became valuable only once Golde and his scientific techniques transformed them into property. The court was concerned about the economic implications of allowing Moore rights to the patent because many universities argued that to allow individuals property in their bodies could stifle the innovation that was promised by the profit motives of this new breed of scientific entrepreneur. Alienating the tissue from the person becomes necessary for the economy to function.

These two cases illuminate the issues discussed in this chapter. In both, parts of a person's body are isolated and transformed into tools. This aspect of these cases shows that the researcher has long been shaped into reductionist engagement with the world, but the ways that the world was instrumentalized were very different. Only in the case of Mo cells were the parts of the body patented and transformed into marketable capital. HeLa cells were envisioned as a resource to be shared by the entire scientific community. Both can lead to injustice to the subjects of reduction, the patients. Both of these modes of engaging the world, the entrepreneurial and the more reductionist, need to be confronted. They are both most directly confronted through attempts to oppose reductionism.

Conclusion

This chapter has advanced the argument of the book by showing that one of the fundamental causes of the problems facing contemporary science and ethics in general is the loss of a teleological vision of the world. Unfortunately, the Aristotelian solution to this issue proposed by MacIntyre, engagement in craft-like communal practices, will not be successful in solving these issues in science. Though training in science does take the requisite form of a craft-like apprenticeship, the daily embodied practice of the contemporary laboratory shapes one in reductionist rather than teleological modes of thought. Thus, these practices serve to lay the foundations for the entrepreneurial vision of science through a vision of the world as commodifiable objects of utility.

To confront these problems, we need a broader theory of practices. Practices need to be understood as capable of shaping a broad variety of character types, not just virtuous ones that embrace teleology. More importantly, we need practices that would enable one to counteract both harmful social structures and the deleterious effects of one's broader daily practices. These characteristics are found in the practice theory developed in the social sciences, especially that of Michel Foucault. It is to this theory that we turn in Chapter 4.

CHAPTER 4

The Practices Shaping the Entrepreneurial Subject

One of the greatest strengths of the craft account of moral formation discussed in the last chapter is that it suggests how values, virtues, and a worldview can emerge organically out of a social practice. Such character formation can become problematic, as with the reductionist cast of mind shaped by daily laboratory practices, but generally it is positive, shaping valuable character traits like temperance, concern for the common good, and a commitment to things that are ends in themselves. Perhaps most importantly for many in contemporary ethics, this craft account helps to assuage concerns over the problem of power. One of the questions that has troubled modern ethics is how to deal with issues of hierarchy, situations in which some people have control over others. While neoliberals address these issues by giving control to the impersonal market, the craft paradigm takes a different route. Crafts do not so much place one person in authority over others as place everyone involved under the authority of the ends and values of the craft. This authority of the craft provides a framework that limits individual control, preventing arbitrary exercises of power. Thus, crafts serve to shape a community of people in a set of dispositions that help to meet the ends of the craft practice, which itself serves to meet the needs of human flourishing.

A very different understanding of social practices, however, emerged from the social sciences, particularly in a strand of practice theory that arose in French social theory in the 1970s.[1] Theorists such as Michel Foucault and Pierre Bourdieu looked less at the coherence of certain craft practices, but instead focused on the techniques employed in daily life and how these influenced the networks of power in society. They examined

[1] Michel Foucault, *Discipline and Punish: The Birth of the Prison* (New York: Pantheon Books, 1977); Pierre Bourdieu, *The Logic of Practice*, trans. Richard Nice (Stanford, CA: Stanford University Press, 1992); Michel de Certeau, *The Practice of Everyday Life* (Berkeley: University of California Press, 2011).

things like the organization of a house, the way hospitals function, and spiritual techniques such as confession. Their conclusion was not that these social practices were places where certain kinds of virtuous dispositions emerge organically to serve ends preset by larger practices. Instead, they found that disparate kinds of practices could be rearranged in various configurations to shape people in very different ways. These practices served to form people into certain kinds of subjects – meaning people with certain kinds of character traits, ways of viewing the world, and frameworks for relating to oneself – that matched a particular ideology or form of organizational functioning. Instead of ways to obviate power, in this view, practices become examples of power at their most insidious, since they do not merely shape action but transform character itself. As we will see, these kinds of analyses help us to make sense of how science shapes the reductionist and the entrepreneurial subject.

Other critics of modernity have addressed the ways that practice shapes behavior by describing problems in managerial power. For someone like MacIntyre, the manager perpetuates a disingenuous manipulation of a subordinate's actions, a form of pure control.[2] Some read Foucault in a similar way, as saying that here everything reduces to the pure power of manipulation. In reality, Foucault's analysis is much more complicated since the task of these practices is to shape subjects who are free, to produce people who freely follow certain ideals. The entrepreneurial manager attempts to transform her employees into mini-entrepreneurs and encourages their activities through the same kind of remuneration in stock options that the entrepreneur receives. The goal is to form people who use their freedom in a certain way. While some of these pressures and practices are clearly external to the person, almost coercively meant to change her, the main thrust of such efforts in a liberal society is to encourage people to take upon themselves certain identities, such as the entrepreneurial identity, and to make efforts to shape themselves in line with that identity. These practices are not merely disciplinary or regulatory practices, but practices of the self. The first two sections of the chapter will discuss these aspects of the use of practice, while showing how this analysis illuminates aspects of the problem of contemporary science.

The chapter will conclude by discussing how the importance of self-directed change in this framework opens space to oppose dominant conceptions of the subject, since it leaves a person the ability to shape

[2] Kelvin Knight, *Aristotelian Philosophy: Ethics and Politics from Aristotle to MacIntyre* (Cambridge: Polity, 2007), 164.

herself into a different kind of subject. Foucault went exactly in this direction by exploring Stoic models of shaping subjectivity, arguing for the importance of ethics. Since many of these Stoic practices of the self were taken up into Christianity, it is also a route that Christian ethics could engage in an attempt to counteract reductionist and entrepreneurial forms of the subject.

Practices That Operate on the Subject

French practice theory is perhaps most famous for its emphasis on power, but theorists like Foucault do not take a naïvely oppositional stance toward power. Power is not just domination or repression.[3] As James Mackey argued, power is a neutral ability to bring about states of affairs, and exists on a spectrum between force and authority.[4] Though power is not exclusively negative, since it is necessary for any kind of social grouping, it is dangerous, and so should always be subject to a critical analysis.[5] In this model, practices function as the technical means that social power uses to shape individual character. Each of these practices or techniques has an independent history, because their function and meaning change over time.[6] For example, the meaning and end of a practice like confession is very different in Christian monasticism than in Freudian psychoanalysis. Part of the reason for these transformations is that smaller practices are integrated into broader technologies of power, each of which has its own vision of the subject.[7] This description of techniques means that for Foucault, as for MacIntyre, the contemporary ethical landscape and

[3] For Foucault, power is merely the attempt to shape another's conduct (Michel Foucault, "The Ethics of the Concern for Self as a Practice of Freedom," in *Ethics, Subjectivity, and Truth*, ed. Paul Rabinow [New York: New Press, 1998], 292). In this form, power is part of all human relationships, operating in situations that most people see as completely legitimate and salutary, such as the teacher–student relationship, love relationships, or political rule (Foucault, "The Ethics of the Concern for Self as a Practice of Freedom," 293–9). These legitimate relations can be perverted: power can be used for control, to prevent any reversibility in influence, or to use others for one's own gain; it can operate through coercion, threats, and spectacular violence; it can seek to covertly shape another's desires and character. For a discussion of Foucault's conceptions of power in relation to Christian sources, see J. Joyce Schuld, *Foucault and Augustine: Reconsidering Power and Love* (Notre Dame, IN: University of Notre Dame Press, 2003).

[4] James Mackey, *Power and Christian Ethics* (Cambridge: Cambridge University Press, 1994), 7.

[5] Michel Foucault, "On the Genealogy of Ethics: An Overview of Work in Progress," in *Ethics, Subjectivity, and Truth*, ed. Paul Rabinow (New York: New Press, 1998), 256.

[6] Michel Foucault, *Society Must Be Defended: Lectures at the Collège de France, 1975–76*, ed. Mauro Bertani and Alessandro Fontana, trans. David Macey (New York: Picador, 2003), 30.

[7] For discussion of Foucault's thought on technology, see Michael C. Behrent, "Foucault and Technology," *History and Technology* 29, no. 1 (2013): 54–104.

language are fragmented. They are a mix of practices and rationalities thrown together despite their independent histories. Unlike MacIntyre, Foucault – as a genealogist – sees this fragmentation as inevitable for any ethical system. Yet, unlike Jeffrey Stout, Foucault does not celebrate this assemblage as an ethical bricolage,[8] since it was formed not through the conscious, rational activity of any individual or community attempting to solve ethical dilemmas but through accidents of history. These assemblages are possibly dangerous, though not necessarily so. It is important to him that we unearth how we arrived at our current historical position, since an understanding of history allows us to think otherwise and find different ways to answer current problems.

Foucault calls the most dominant shape that power takes today biopower. It uses both disciplinary practices and regulatory practices as external measures to shape the individual subject.[9] Biopower attempts to maximize the productive capacities of citizens by acting on their bodies, on their very biology, both as individuals and at the level of the population. In order to manipulate these productive capacities well, this system of power utilized a highly mechanistic, reductionist understanding of the body.[10] Given a new focus on efficiency in early modern Europe in areas such as professional standing armies and capitalist modes of production, new practices came into being that break down bodily action into the smallest component maneuvers and reassemble them in the most efficient way possible. They treated the body as a machine to be optimized, and they trained the body to act efficiently. These practices did not arise all at once in a centrally planned operation. Instead, they arose one at a time, beginning in the army, but becoming generalized to many other kinds of institutions, such as the hospital, the factory, the school, and, of course, the prison. Individual techniques tended to come together at these large enclosed facilities and thereby reproduce the regime of disciplinary technology across society.

The analysis of performance that occurs in discipline is not juridical in the form of whether one violates a law or not. Instead, the individual is measured against a norm of performance,

> [a]n optimal model that is constructed in terms of a certain result, and the operation of disciplinary normalization consists in trying to get people,

[8] Jeffrey Stout, *Ethics after Babel: The Languages of Morals and Their Discontents* (Boston: Beacon Press, 1988), 74–7.

[9] Michel Foucault, *The History of Sexuality*, vol. 1 (New York: Pantheon Books, 1978), 136–45.

[10] Foucault, *Discipline and Punish*, 135–228.

> movements, and actions to conform to this model, the normal being precisely that which can conform to this norm, and the abnormal that which is incapable of conforming to the norm.[11]

Such a norm would be the ideal speed of manufacture, performance in a class, or disease progression. While these practices work on and train the body, the body itself is not their object, but "an instrument or intermediary."[12] They act on the soul, and the form of the contemporary soul, "psyche, subjectivity, personality, consciousness, etc.," is brought into being through these practices. Through bodily practice, the individual is trained to act efficiently, to act mechanically, a habit that suffuses her entire life and subjectivity. Thus, the rise of a mechanistic view in science interacted with other cultural trends, and the reductionist techniques discussed in the last chapter were not limited to science or medicine. Instead, these take part in a larger worldview.

How these practices work to shape dispositions can be better understood by looking at Pierre Bourdieu's concept of habitus. Bourdieu also emphasizes the importance of bodily training in forming subjectivity. Practical sense and knowledge are embodied in motor schemas and bodily automatisms that are similar to Polanyi's tacit knowledge. The social order uses the body as a depository "of deferred thoughts that can be triggered off at a distance in space and time by the simple effect of re-placing the body in an overall posture which *recalls* the associated thoughts and feelings."[13] If a child is taught to kneel during prayer, kneeling later in life will bring back a whole set of affective and practical dispositions related to prayer. Learning to sit at a desk at school recalls a similar set of dispositions when one is again seated at a desk to study. Values are "made body" by "an implicit pedagogy which can instill a whole cosmology" through seemingly insignificant forms.[14] Embodied relation to space, posture, and action is of central ethical import for the formation of worldview and affect.

This attention to the relationship between embodied practice and subjectivity, between social organization and moral formation, adds to the analysis of character formation in MacIntyre and Polanyi, especially

[11] Michel Foucault, *Security, Territory, Population: Lectures at the Collège de France, 1977–78*, ed. Michel Senellart, trans. Graham Burchell (New York: Palgrave Macmillan, 2007), 57.

[12] Foucault, *Discipline and Punish*, 11.

[13] Bourdieu, *The Logic of Practice*, 69. Emphasis in the original.

[14] Bourdieu, *The Logic of Practice*, 69.

in the consideration of a scientific education. This kind of embodied training is central to an undergraduate's first weeks and months in the laboratory. The postdoctoral fellow or graduate student who trains an undergraduate assistant trains her first in the dullest and most monotonous tasks of the laboratory. The training consists in a day spent making solutions and gels: following a recipe in the most minute detail, pouring out the exact weight of ingredients, using the pH meter to perform a delicate titration of dangerous acid. Perhaps the undergraduate will be set to work sectioning tissue: learning to embed the sample in the exact position necessary before the paraffin hardens, sitting in front of a microtome for hours on end, learning to control her arm so that the blade stroke is steady and the tissue is not damaged, gently using a paintbrush to lift six-micron-thick sections from a water bath onto the slide without the tissue falling out. These are arts in one sense, skills that the researcher will take pride in once they are mastered, but these processes also involve learning to adapt one's body to instruments, to make one's body an instrument. They of course teach virtues that are important in science and the rest of life – precision, care, and patience, to name a few – yet they also train one in a way of viewing animal tissue as open to fragmentation. They adapt one to the machines in the laboratory, teaching the student to move precisely like these machines. Moreover, they serve as a test. Can this undergraduate learn to be careful, to master this skill, to meet the norm of performance? If not, the student will not receive the necessary letters of recommendation and will not have a career in science.

Many types of research also require the training of affect. The first days spent with any animal system is a training in overcoming initial physical rejections of working with the system. My first week working in a chick laboratory involved the process of learning to competently perform microsurgeries. In reality, I did not learn to perfect microsurgeries until much later, since the first day working with actual chicken eggs involved developing the skill necessary to open a window in the eggshell by tapping holes in each end of the egg, using a hypodermic needle to slowly withdraw just enough egg white to cause the chick embryo to fall away from the shell but not enough to kill the embryo, and then cutting a large hole in the top of the eggshell with scissors. If performed with too much force or at the wrong point, any one of these steps can cause the egg to crack. With my untrained hands, I seemed to crack one egg for every one that I properly windowed. By the end of an afternoon, my jeans and shirt were covered in egg. By the end of the week, I gagged at the smell of egg, to the point of not being able to eat eggs for a few weeks.

Similarly, on my second day working in a mouse laboratory, I was assigned the mouse cull, a necessary task in any genetics laboratory. In order to avoid unnecessarily maintaining extra mice, which would be both a waste of resources and stressful for the animals, researchers kill unneeded animals by placing them in carbon dioxide, which anaesthetizes them as it kills them. It is a long afternoon of killing if it is your first time, an afternoon that requires overcoming discomfort with such activity. Because of this innate discomfort with killing, it is also a test. If one cannot overcome one's dislike of the task, then one cannot continue in the laboratory, so new students are frequently assigned such tasks in their first weeks in a laboratory to see if they will be able to adjust themselves to these requirements. Such activities distance the scientist from his research tools, like animals, preventing him from becoming too affectively engaged with them as whole organisms. The trainee overcomes not just intellectual impulses but physical reactions to sights and smells. This kind of embodied discipline is important in the formation of scientific subjectivity, shaping the trainee to see organisms as tools.

The practices that channel contemporary power do not just act on the body of the individual, but act in a second way through regulatory tools applied to the population as a whole. With the rise of vital statistics such as birth rates, death rates, epidemic disease frequency, etc., there appeared a set of regularities that bureaucrats cannot see by looking at individuals.[15] A new object came under consideration, that of the population. Political economists and public health specialists argued that the population follows its own laws and natural processes that government must respect.[16] It cannot be forced to abide by just any norm that one arbitrarily chooses because it has its own laws, but, as a statistical distribution, its normal curve can be adjusted. Through this focus on the statistical normal curve, one can find areas that do not have the same distribution as the rest of society in terms of some variable and try to act in order to force that distribution back into line with the more general distribution of the population as a whole.[17] If obesity rates rise in one area, then the government might change zoning laws to eradicate food deserts, launch educational campaigns to encourage exercise, or ban large sodas. In all of these cases, the method of power is regulation that attempts to act on people's choices in a general manner in order to modify a statistical irregularity, in

[15] Foucault, *Society Must Be Defended*, 243.
[16] Foucault, *Security, Territory, Population*, 348–54.
[17] Foucault, *Security, Territory, Population*, 60–3.

this case a rise in the rate of obesity in a distinct sector of the population. It uses educational campaigns, statistical knowledge, pension schemes, risk avoidance measures, and insurance as its tools, rather than the manipulation of the individual body.

Science is also managed through such technologies of regulation that attempt to change the behavior of a large group of specialists. The changes to patent law, encouragement of academic patents, and the financial innovations of the biotech industry that were discussed in Chapter 2 are all broad alterations to the rules of the game that attempt to change how scientists as a group view themselves and act. These changes attempted to shape the distribution of how many basic research discoveries are transformed into intellectual property and how many scientists devote themselves to practical applications. So far, it has been a remarkably successful deployment of power to alter behavior.

Practices of the Self and the Entrepreneur

Contemporary discussions of biopower can seem incredibly bleak and are easily taken to be total critique. Humans are viewed as machine-like tools with bodies, desires, and worldviews shaped in institutional settings. Large-scale regulations shape behavior at the level of the population. Contemporary politics contains the potential for recurrences of aspects of Nazism or Stalinism. Foucault seemed to recognize these problems in his thought and soon denounced inflationary critiques of the state, critiques that see an intrinsic dynamism of the social welfare state toward totalitarianism: "An analysis of social security and the administrative apparatus on which it rests ends up, via some slippage and thanks to some plays on words, referring us to the analysis of the concentration camps."[18] To address these problems, he also turned to focus on the importance of freedom in how the

[18] Foucault, *Society Must Be Defended*, 187–8. He took some of this care to prevent illegitimate inflation in his earlier critique of liberal power, by specifying common rationalities and techniques rather than making a blanket denunciation. However, many writers see a fundamental change in Foucault's political thought between his 1976 and 1978 lecture series, which included both a rejection of much of politics on the left and a concern with the rise of neoliberalism (Árpád Szakolczai, *Max Weber and Michel Foucault: Parallel Life-Works* (New York: Routledge, 1998); Eric Paras, *Foucault 2.0: Beyond Power and Knowledge* (New York: Other Press, 2006); Michael Behrent, "Liberalism without Humanism: Michel Foucault and the Free-Market Creed, 1976–1979," *Modern Intellectual History* 6, no. 3 (2009): 539–68; Alain Beaulieu, "Towards a Liberal Utopia: The Connection between Foucault's Reporting on the Iranian Revolution and the Ethical Turn," *Philosophy and Social Criticism* 36, no. 7 (2010): 801–18). The reasons for this shift are complex. He grew increasingly disillusioned with the Maoists and other strains of post-1968 political thought, while becoming more interested in the *nouveaux philosophes*, like Andre Glucksmann

contemporary subject, especially the entrepreneurial subject, is formed. His notion of power and, he argued, society's deployment of power increasingly included the governance of conduct or, in playing on the meanings of the French word *conduit*, the conduct of conduct. "To govern ... is to structure the possible field of action of others."[19]

The Christian pastorate provides one model of how one governs freedom. The Church is "an institution that claims to govern men in their daily life on the grounds of leading them to eternal life in the other world, and to do this not only on the scale of a definite group, ... but of the whole of humanity."[20] The shepherd has many functions, in that he must protect the sheep, make sure they have food and water, and track down the lost. Similarly, the pastor must keep an eye on the entire life of each of his congregants to check the sins that might arise in any area of existence. As the parable of the lost sheep demonstrates, the pastor must ensure that the Church as a whole is led to salvation, but also has a responsibility for each individual member, risking the safety of the flock to seek him out.[21] In these two foci of attention lie the precursors to biopower's twin forms of regulation of the population and discipline of the individual. Further, pastoral power is neither political nor juridical, but, rather, therapeutic.[22] Because they are different forms of power, pastoral power can coexist with juridical sovereignty, as they did in the medieval period.[23]

In the early modern period, the locus of this pastoral control began to shift to the state, which saw that it had to encourage its subjects to behave in certain kinds of ways to gain military and economic strength. It did this in part through the coercive means of disciplinary and regulatory practices discussed above. At same time, it also slowly began to attempt to shape how people used their freedom, much in the way that pastors tried to encourage their congregations to shape themselves into people dedicated

and Bernard-Henri Levy, who embraced neoliberal ideas. His disappointing engagement with the Iranian Revolution led him to increasingly engage in human rights advocacy. His intellectual interlocutors changed from Gilles Deleuze to his classicist colleagues at the College de France like Paul Veyne and Pierre Hadot and interlocutors at UC Berkeley, such as Paul Rabinow, Hubert Dreyfus, and Peter Brown.

19 Michel Foucault, "The Subject and Power," in *Power*, ed. James Faubion (New York: New Press, 2001), 341.

20 Foucault, *Security, Territory, Population*, 145.

21 Foucault, *Security, Territory, Population*, 168–9.

22 Foucault, *Security, Territory, Population*, 174.

23 Juridical forms of power arise in the Church of course, which Foucault and others like John Mahoney, *The Making of Moral Theology: A Study of the Roman Catholic Tradition* (New York: Oxford University Press, 1989), 1–36, have noted in regard to practices of confession. The increasing juridification of the Church opened greater possibilities for conflict between realms.

to God. Foucault calls the modern secular conduct of conduct through control of populations governmentality.[24] Since becoming a secular project, governmentality has undergone many shifts, but today its major vehicle is neoliberalism, which attempts to form individuals as entrepreneurs acting in a competitive marketplace. Foucault sees current conflicts over religion as not fundamentally between Church and state or pastorate and law, but between pastoral power and contemporary governmentality because of their similarity.[25]

Governmentality operates by targeting mindsets and dispositions through trainings and ideals. The problems facing society in many instances are identified as originating in wrong habits of thinking, which makes changing attitudes a fundamental task of the government.[26] To change these attitudes, one not only uses the coercive powers of discipline and regulation, but one tries to prompt the individual to take on a new ideal and shape herself in line with the ideal. One of the primary ways that government and business attempt to shape subjectivity is through trainings that propose a new way of being in the world. Emily Martin discussed how throughout the 1980s and 1990s businesses encouraged their workers to become flexible.[27] The key to this flexibility was the ability to actively tolerate risk, which is important to an entrepreneurial subjectivity, as we have seen. This training was not merely a series of dry lectures; it included embodied engagement with risk through team-building exercises on ropes courses, for example. These were exciting trainings that motivated workers to think of themselves as flexible, to consider risk as something to be embraced rather than avoided. Such an attitude was beneficial in a corporate landscape in which workers faced massive downsizings and many jobs ceased to be permanent positions with benefits and pensions but instead became precarious freelance contract work. Such trainings sought to make these changes seem like the realization of an ideal rather than the loss of older forms of job security. By shaping the workers' ideals, such a condition could become the realization of their ideals.

Barbara Ehrenreich has chronicled how such measures were continued into the early 2000s through the slightly different means of motivational

[24] Foucault, *Security, Territory, Population*, 108. For an introduction to governmentality, see Michel Foucault et al., *The Foucault Effect: Studies in Governmentality* (Chicago: University of Chicago Press, 1991); Mitchell Dean, *Governmentality: Power and Rule in Modern Society* (Thousand Oaks, CA: Sage Publications, 1999).

[25] Foucault, *Security, Territory, Population*, 191.

[26] For a discussion of how this mode of engaging mindsets was implemented under Margaret Thatcher, see Christopher Payne, *The Consumer, Credit and Neoliberalism: Governing the Modern Economy* (New York: Routledge, 2012), 82–4.

[27] Emily Martin, *Flexible Bodies* (Boston: Beacon Press, 1995).

speaking and positive psychology.[28] In these cases, many times in the aftermath of layoffs, motivational speakers encouraged workers to focus on the positive, to be excited about risk, and to not fear the negative. Moreover, through positive psychology, trainings give individuals the ability to shape their own dispositions through sets of exercises. A recent World Bank study suggested that psychology-based training focused on forming an entrepreneurial mindset was more effective than traditional business trainings in encouraging successful business in West Africa.[29] Such trainings use many of the same techniques encouraged by ancient philosophers and developed in the religious devotional exercises to manipulate one's own thoughts that we will discuss in Chapter 5.[30] However, they do so in order to encourage one to shape a subjectivity and a series of dispositions that are distinct from more traditional religious aims. Sam Binkley suggested that these kinds of exercises in positive psychology are closely tied to shaping an entrepreneurial subjectivity. For example, they tend to encourage one to make things in this world, including other people, into resources for one's own quest for happiness.[31] They train people to consider the relationship between activities and happiness from the perspective of a cost–benefit analysis, asking people to analyze their efforts in terms of what will take the least time to yield the most happiness.[32] Thus, one is encouraged to realize an ideal in oneself through sustained work on the self.

While not as highly developed in science, there is a push for increased inclusion of entrepreneurial training in the education of scientists, as discussed in the Bioeconomy Blueprint. The NSF's I-Corps and similar programs at many universities and state agencies seek to mentor scientists along the path to entrepreneurship. Currently, in the case of science, most of the encouragement for taking upon oneself an entrepreneurial identity occurs at the level of the laboratory, in discussions of entrepreneurial scientists in popular publications, and through the mentors who form the disciplinary ideals. However, it is likely that more thorough entrepreneurial training paradigms soon will come into being for scientists as they have for many other fields.

[28] Barbara Ehrenreich, *Bright-Sided: How Positive Thinking Is Undermining America* (New York: Picador, 2010).

[29] Francisco Campos et al., "Teaching Personal Initiative Beats Traditional Training in Boosting Small Business in West Africa," *Science* 357, no. 6357 (September 22, 2017): 1287–90.

[30] Sam Binkley, *Happiness as Enterprise: An Essay on Neoliberal Life* (Albany: SUNY Press, 2014), 30.

[31] Binkley, *Happiness as Enterprise*, 35.

[32] Binkley, *Happiness as Enterprise*, 33.

Avoiding the Entrepreneurial Imperative

If an entrepreneurial character that views the world and other humans in a reductive manner is encouraged not only by disciplinary and regulatory practices but also by individuals' own acceptance of it as an ideal, it leaves open the question as to how one could change this social ideal or even prevent oneself from becoming enamored by it. While many commentators note that Foucault was intrigued by neoliberalism,[33] given its focus on freedom, this growth of entrepreneurial governmentality posed at least two problems, even from his standpoint. First, these mechanisms of governance do have coercive aspects, aspects that share a relationship with the totalitarian regimes of the twentieth century. Second, they are, in the end, examples of government expertise imposing a form of subjectivity on individuals.[34] Broad movements of social protest show that people want to relate neither to themselves as entrepreneurs nor to others in the mode of competition. One possible form of pushing back against these forms of subjectivity embraced by many Christian ethicists are counterpractices, practices that resist the dominant social trends.[35] Yet, because governmentality shapes subjectivity, if one does not begin by reshaping one's subjectivity, then overt political resistance may merely reinforce neoliberal forms of subjectivity. This problem leads to Foucault's work on ethics, suggesting that actively caring for oneself, shaping one's own subjectivity in an alternative manner, is necessary. This possibility is left open by the way that governmentality operates.

Since power is an attempt to shape another's conduct, it requires the possibility that the other is capable of action, has some degree of freedom to act in an unexpected way.[36] In some technologies of power, such as chattel slavery, that freedom may be so constrained that the possibility of resistance may be reduced to violence. But since contemporary governmentality works by shaping the subject of freedom, it gives a great degree

[33] He even seemed to be attracted to specific neoliberal policy proposals such as the negative income tax, similar to the earned income tax credit, because it reduced the disciplinary power of the welfare system. See Foucault, *The Birth of Biopolitics*, 203f.; Behrent, "Liberalism without Humanism: Michel Foucault and the Free-Market Creed, 1976–1979"; Paras, *Foucault 2.0*.

[34] Foucault, "The Subject and Power," 330.

[35] Gerald McKenny, *To Relieve the Human Condition: Bioethics, Technology, and the Body* (Albany: State University of New York Press, 1997); Jonathan Tran, *Foucault and Theology* (New York: Bloomsbury T&T Clark, 2011).

[36] Foucault, "The Subject and Power," 342; Foucault, "The Ethics of the Concern for Self as a Practice of Freedom," 292.

of freedom of action. While contemporary forms of power may work at a deeper level than earlier forms by more directly attempting to shape subjectivity, by that very fact they allow more possibilities of resistance.

A degree of freedom is not unlimited choice, however. Resistance will always occur within the rationality of the technology of power that it is resisting. Individuals cannot escape the terms of contemporary paradigms of knowledge, since contemporary power relations form one's ways of thinking. That is why questions of truth are always an important field of resistance. Since knowledge frequently serves to reinforce power, power can be combatted by counterknowledges.[37] For example, if a regime of disciplinary eugenics sought to sterilize prisoners and those with below-average IQ scores by claiming that their children would possess the same traits, then one way to resist such sterilizations would be to challenge the idea that such traits were strictly genetically determined. One can, as many did, point to social and environmental factors. In terms of entrepreneurial science, one can use data such as those presented in the first chapter to question whether a more entrepreneurial science truly achieves the goals of societal benefit that it claims to seek. As Weber put it, one can point out inconvenient facts.

More commonly recognized, one can confront certain disciplinary practices with counterpractices. For example, one form of resistance to biopower involves returning to the language of rights of juridical sovereignty. Resistance adjusts this discourse to become positive rights to life. If biopower attempts to shape the mechanisms of life for the state's or civil society's needs, then individuals and groups respond by legally fighting for rights to aspects of life that they see as essential.[38] This could be more stable working conditions or less pressure from competition.

[37] Foucault, "The Ethics of the Concern for Self as a Practice of Freedom," 295. Truth is very important in Foucault's politics, even if his conception of it is complex. These notions of truth and their relation to theological understandings of truth will be explored further in Chapters 5 and 7.

[38] Foucault, *The History of Sexuality*, vol. 1: 145. Foucault's notions of rights to life are not the same as one might find in many Christian congregations like the Catholic Church. For Foucault, these are moral notions specific to a particular discourse, not timeless qualities of the human, since for him all notions are tied to particular formations of knowledge and power, particular histories, and particular practices. Merely because they are not a timeless essence does not mean that they do not exist, though. Like madness or sexuality, they are a product of knowledge and practice at a particular moment. Even at the level of specific rights there is a great difference. While some overlap, like full employment and opposition to eugenics, others do not (Foucault enthusiastically supported abortion and suicide). That does not mean that one could not take up his analytic tools for different ends, though. John Paul II saw the prolife movement as fighting the totalitarian impulses of liberal society that Foucault analyzes under the name of biopower, so these tools can be deployed for other struggles.

Yet regimes of governmentality attempt to shape subjectivity in a way that is not amenable to standard methods of opposition. Protests, legal challenges, new forms of community, etc. all have the potential to merely reinforce a particular kind of subject of freedom when used against governmentality. Merely engaging in rights talk risks devolving into a libertarianism that reinforces laissez-faire economics. Opposing coercive disciplinary eugenics programs of sterilization may merely privatize eugenics to those enterprising families who can afford to maximize their human capital. As many science studies scholars have found to their dismay, challenging discursive formations of power/knowledge risks playing into the hands of corporate interests who, for example, use such analyses to undermine research indicating anthropogenic climate change.[39] As MacIntyre notes, rights can combine with utility to merely reinforce existing power relations.

What is needed instead is a shift in subjectivity. One must be transformed in how one uses one's freedom. While there is much to be decried in contemporary liberalism, it offers a space of freedom for acting on oneself, which means that ethics can be a way to confront governmentality's influence on subjectivity. By ethics, social theorists following Foucault refer to the reflected relation of oneself to one's self.[40] It is the elaboration of a distinct style of life, the pursuit of what Hellenistic ethics called the care of the self. It is a conscious taking in hand of one's moral formation, not merely leaving it up to surrounding social forces.

Such an ethics is not a rejection of moral rules; at least it does not have to be. Like MacIntyre, Foucault decries the focus of philosophy and social science on formal moral norms, but, again like MacIntyre, his work is meant to better articulate the connections between rules of conduct and

[39] Bruno Latour, "Why Has Critique Run Out of Steam? From Matters of Fact to Matters of Concern," *Critical Inquiry* 39 (2004): 225–48; Naomi Oreskes and Erik Conway, *Merchants of Doubt: How a Handful of Scientists Obscured the Truth on Issues from Tobacco Smoke to Global Warming* (New York: Bloomsbury Press, 2010). For further discussion, see Chapter 6.

[40] Michel Foucault, *The Use of Pleasure* (New York: Vintage Books, 1985), 6. For other social theoretical accounts that follow Foucault's discussion of ethics, many of which discuss it in terms of neoliberalism and religious ethics, see Paul Rabinow, *Anthropos Today: Reflections on Modern Equipment* (Princeton, NJ: Princeton University Press, 2003); Joel Robbins, *Becoming Sinners: Christianity and Moral Torment in a Papua New Guinea Society* (Berkeley: University of California Press, 2004); Saba Mahmood, *Politics of Piety: The Islamic Revival and the Feminist Subject*, 2nd revised edition (Princeton, NJ: Princeton University Press, 2005); Jarrett Zigon, *"HIV Is God's Blessing": Rehabilitating Morality in Neoliberal Russia* (Berkeley: University of California Press, 2010); James Laidlaw, *The Subject of Virtue: An Anthropology of Ethics and Freedom* (New York: Cambridge University Press, 2014); China Scherz, *Having People, Having Heart* (Chicago: University of Chicago Press, 2014).

character. As with Kant and Hegel, he separates morality from ethics. However, his reason for doing so is that while rules of conduct remain fairly constant, the way that people relate to those rules changes radically from one philosophical, religious, or cultural tradition to another.

He separates this relationship of the self to rules of conduct into four categories.[41] First, there is the question of the ethical substance. What is it that one takes to be important for ethical development? Is it a Stoic focus on reason, coherence, and passions; Augustinian desire; or disciplinary efficiency? Second, there is the issue of the mode of subjection. How does one relate to rules? Is it Christian obedience to God, Kantian duty, or Stoic self-mastery? Third, there is the ethical *telos*. What is the flourishing life and good subjectivity at which one aims? Finally, there is the ethical work required to reach the desired subjectivity. In MacIntyre's terms, it is the way one gets from man-as-he-happens-to-be to man-as-he-could-be-if-he-realized-his-essential-nature.[42]

It is this fourth point that Foucault's last works elaborated, attempting to take the practices of self-shaping that are used in current forms of governmentality and use them to shape a different kind of self. Following Pierre Hadot's studies on spiritual practices,[43] Foucault turned to Hellenistic ethics, especially Stoic ethics, to develop his version of ethics. These systems were attractive because they showed models of ethics that focused on the form and style of life, and he thought that the social position of the Stoic philosophers was surprisingly similar to that of modern subjects. Of course, one cannot resurrect an ethic of another time to serve us now, but it can be useful for helping us to conceptualize a new ethic for today. It is this model of the care of the self that offers a chance to oppose problematic models of neoliberal subjectivity. In Chapter 5, I will compare this Stoic response to the problems facing contemporary science with what would result from the Aristotelian response that MacIntyre would encourage.

41 Foucault, *The Use of Pleasure*, 26–8.

42 Alasdair MacIntyre, *After Virtue: A Study in Moral Theory*, 2nd edition (Notre Dame, IN: University of Notre Dame Press, 1984), 52.

43 Pierre Hadot, *Philosophy as a Way of Life: Spiritual Exercises from Socrates to Foucault*, trans. Arnold I Davidson (Cambridge: Blackwell, 1995).

CHAPTER 5

Reshaping the Entrepreneurial Subject of Science

The argument presented so far suggests that contemporary moral problems can be solved only if individuals' lives take a different form. People must cease to engage the world as material for use or sale, acting as agents of manipulation and competition, and instead undergo a different kind of moral formation. Disconcertingly, the philosophers discussed in the last two chapters prescribed radically different strategies for enacting this change. In MacIntyre and those he has influenced, change can come only through engagement in alternative forms of social practice like craft, while in Foucault, though not denying the value of such counterpractices, the individual must also undertake conscious work on his self.

This chapter traces these differences in approach to the different schools, or the different periods, of Greek philosophy from which each of these thinkers draws his theory of moral formation. MacIntyre argues that a change in political structures must come before ethical transformation because one becomes virtuous by living within a small community devoted to virtuous practice, a model that largely derives from Aristotle's argument that virtues arise from habituation in a well-structured *polis*. In contrast, Foucault sees personal transformation as both necessary for and partly constitutive of political transformation. Drawing on Stoicism, Foucault argues that ethical transformation occurs through the subject's work on the self, work that consists of practices of the self, including meditative and ascetic techniques.[1] These techniques gradually rework one's subjectivity,

[1] One can find practices of the self in Christian Aristotelians like Thomas Aquinas (Matthew Kruger, "Aquinas, Hadot, and Spiritual Exercises," *New Blackfriars* 98, no. 1076 [July 1, 2017]: 414–26). Aquinas inherited this Stoic tradition through its influence on Patristic authors and the monastic tradition of religious practice (Michel Spanneut, *Le Stoïcisme des Pères de l'église, de Clément de Rome à Clément d'Alexandrie*, Patristica Sorbonensia 1 (Paris: Éditions du Seuil, 1957); Marcia L. Colish, *The Stoic Tradition from Antiquity to the Early Middle Ages*, 2 vols. (New York: E. J. Brill, 1990); Richard Sorabji, *Emotion and Peace of Mind: From Stoic Agitation to Christian Temptation* (New York: Oxford University Press, 2000). The Aristotelian/Stoic contrast that I am making is of ideal types, which are found in complex forms in living traditions.

allowing one to become a subject of truth in line with an ethical tradition. In these two authors, we see two different models of moral formation drawn from Antiquity applied to the problem of the modern subject.

The goal of this chapter is to evaluate which of these solutions better addresses the problems of contemporary science. Despite the predominance of the emphasis on social practices in much of religious ethics, an engagement with practices of the self better coheres with much of Christian spirituality, which makes use of devotional practices to help the individual believer to live as a wayfarer in a fallen world. Such a focus on spiritual practices also provides a better model for the development of the Christian scientist in contemporary life because it recognizes that the scientist must engage the pluralist ethical influences of the cosmopolitan scientific community. This latter model does not deny the importance of social contributions and social practices to moral formation, but it adds an important emphasis on the role of individual initiative in forming character. Many of the common criticisms of Stoic ethics do not reflect an accurate description of Stoic teaching. To show this unexpected congruence of Stoic ethics with science as a Christian vocation, I will draw on Antonin Sertillanges's portrait of the intellectual life, which has formal similarities to Foucault's. Yet this need for practices of self-formation is not limited merely to the Christian, since secular, even atheist, accounts of science as a vocation such as that found in Max Weber also emphasize its ascetic character. In this chapter, I develop the general analysis of moral formation through practices of the self as a productive model for counteracting the contemporary encouragement of entrepreneurialism and reductionism.

Aristotelian Habituation in Small Community

As discussed in Chapter 3, one way of understanding moral formation sees virtue as formed in the practice of a *techne*. Yet merely engaging in an isolated craft is not enough to ensure the development of virtue. First, the institutions supporting a *techne* could corrupt it either by putting its proper products to bad use or by diverting the *techne* from its intrinsic goods to external goods. For example, the skilled railway worker could put its proper product to bad use by aiding the transport of deportees to Auschwitz.[2] Similarly, mere devotion to science is not virtuous if the

[2] Alasdair MacIntyre, "Social Structures and Their Threats to Moral Agency," *Philosophy* 74, no. 3 (1999): 312.

research produces biological weapons. Second, one must engage in many areas of life and make choices between the goods instantiated in each area. Training in one *techne* will not provide the larger hierarchy of goods necessary for rationally making decisions. The virtuous person needs a community with a common narrative and tradition that will provide an ordered space for all the goods of the various *technai*.[3]

In MacIntyre's eyes, because of its poor philosophical foundation and its bureaucratic nature, the modern state cannot form such a community, an opinion that puts him at odds, as he stresses, with most communitarian thinkers.[4] Only communities set apart from the modern managerial state encourage virtuous character. Over his career, he has developed many models of such communities, starting with isolated contemporary analogs of St. Benedict's monasteries in *After Virtue*.[5] More recent examples have not been quite as separatist, looking to Plato's philosophical community that existed in the *polis* but not as a part of the *polis*.[6] He also discusses small civil society groupings that exist within the state, participate in some of the benefits of the state (aid for people with disabilities, for example), but keep themselves aloof from identification with the state.[7] Such communities "will have a double attitude to the nation-state. They will recognize that it is an ineliminable feature of the contemporary landscape and they will not despise the resources that it affords . . . But they will also recognize that the modern state cannot provide a political framework . . . to achieve the common good."[8] His critics note that his concrete examples are vague, and they frequently question whether the communities he seeks exist.[9] He does point to some possible concrete forms of these virtuous communities, such as relatively isolated fishing villages organized around the *techne* of fishing that still possess an older hierarchical system of goods.[10] He also thinks that the Catholic Church could form such a

[3] Alasdair MacIntyre, *After Virtue: A Study in Moral Theory*, 2nd edition (Notre Dame, IN: University of Notre Dame Press, 1984), 204–25.

[4] Alasdair MacIntyre, *Dependent Rational Animals: Why Human Beings Need the Virtues* (Chicago: Open Court, 1999), 142.

[5] MacIntyre, *After Virtue*, 263.

[6] Alasdair MacIntyre, *Whose Justice? Which Rationality?* (Notre Dame, IN: University of Notre Dame Press, 1988), 99.

[7] MacIntyre, *Dependent Rational Animals*, 135.

[8] MacIntyre, *Dependent Rational Animals*, 133.

[9] E.g., David Fergusson, *Community, Liberalism and Christian Ethics* (New York: Cambridge University Press, 1999), 126–30.

[10] Fergusson, *Community, Liberalism and Christian Ethics*, 143; Alasdair MacIntyre, *Ethics in the Conflicts of Modernity: An Essay on Desire, Practical Reasoning, and Narrative* (New York: Cambridge University Press, 2016), 178.

community, especially given its network of educational institutions.[11] In all the cases that he presents, though, such communities are relatively isolated from the corrupting effects of modernity.

Keith Breen sees this solution as problematic because it seems to require bad faith in the virtuous small community's engagement with the wider society.[12] In MacIntyre's model, the small community relies on the state for the protection of rights, for services, for emergency relief, and for the resources to support those with disabilities. Yet despite the necessity of and benefits received from the state, the community feels no obligation to support this wider society beyond the kinds of participation that can be coerced, acting as a group of free riders. Moreover, this scheme seems to betray the Aristotelian impulse to find the good in existing forms. The state cannot be completely fallen for MacIntyre since he supports aspects of the state such as the Americans with Disabilities Act and the Marine Corps.[13] Instead, MacIntyre valorizes marginal figures, for which he earlier criticized Herbert Marcuse.[14] In Marcuse's case, it was the marginal figures of avant-garde artists that were to bring salvation, while in MacIntyre's, it is the marginal fishing village that can save us. This characterization is somewhat unfair because it ignores MacIntyre's engagement with the Catholic Church and its educational system. Still, the characterization makes a point, since he never explains how the Church corresponds to his small virtuous communities. This problem of whether such a virtuous Church exists is also found in theological affirmations of similar solutions, such as that of Stanley Hauerwas.[15]

The small isolated community is an especially poor solution for a practice like biology or any other contemporary field of science. A scientific discipline forms something like its own community. As the discussion of the correspondence between scientific and ethical formation in Chapter 3 demonstrated, the virtues and worldview of the scientist are deeply shaped by her scientific education. Science requires an induction into a tradition by a master, through which one learns how to think, experience, and act. Further, it embeds one in a cosmopolitan scientific community. Finally, it

[11] Alasdair MacIntyre, "An Interview with Cogito," in *The MacIntyre Reader*, ed. Kelvin Knight (Notre Dame, IN: University of Notre Dame Press, 1998), 267.

[12] Keith Breen, "The State, Compartmentalization and the Turn to Local Community: A Critique of the Political Thought of Alasdair MacIntyre," *The European Legacy* 10, no. 5 (August 2005): 495.

[13] Breen, "The State, Compartmentalization and the Turn to Local Community," 494.

[14] Keith Breen, "Alasdair MacIntyre and the Hope for a Politics of Virtuous Acknowledged Dependence," *Contemporary Political Theory* 1, no. 2 (2002): 198; Alasdair MacIntyre, *Herbert Marcuse: An Exposition and a Polemic* (New York: Viking Press, 1970).

[15] Fergusson, *Community, Liberalism and Christian Ethics*, 48–79.

is a formation that is deeply intertwined with the worldview of technocratic modernity. Thus, this individual formation in the tradition of science must escape the boundaries of the moral formation of small communities of virtuous practice. From a theoretical perspective, biology would be difficult to integrate into the envisioned small communities since the biologist must remain engaged with these broader institutions of the scientific community. More importantly, this individual formation in the daily practice of science, shaping the trainee in habits of reductionism as it does, runs counter to the teleological worldview these small communities should ideally instill.

From a more pragmatic perspective based on the contingent form of contemporary science, the practice of research would be impossible to sustain without engaging the resources that corporate biotechnology or government funding makes available. Such external goods are critical. Advances in scientific knowledge require extremely expensive technologies. Physics requires huge, multinational particle accelerators to test its theories. Contemporary genetics would be unimaginable without the billions of dollars spent on the Human Genome Project and other sequencing efforts, or the vast number of expensive lines of transgenic mice. Cell biology requires sophisticated imaging systems whose combination of laser microscopy, super-cooled cameras, and image processing software can cost nearly a million dollars. To seek new tools to manipulate cell systems, scientists need libraries of small molecules from pharmaceutical companies. Biotechnology companies produce many essential research tools like polymerase chain reaction (PCR), which allows the easy amplification of a DNA sequence.[16] The growing study of the microbiome, which investigates the ecology of bacteria within the human body, requires vast budgets for sequencing microbial genomes and the expensive and difficult task of raising model organisms in sterile environments. These elements are essential to the practice of science.

A scientist can access these resources only by playing the game of contemporary entrepreneurial science. To gain tenure, she must produce papers and secure grants. To procure grants, a project may have to promise broad societal benefits. Translating a concept into reality may require venture capital or collaboration with a pharmaceutical corporation. One could say that the entire structure of science should be changed, but such changes take a long time to enact and need individuals whose characters

[16] See Paul Rabinow, *Making PCR: A Story of Biotechnology* (Chicago: University of Chicago Press, 1996).

are properly formed to enact them. To remain a practicing scientist among the imperfect structures that exist now and in the foreseeable future, and to support those who depend on her for training, the scientist must engage in aspects of this entrepreneurial model.

Of course, this problem is not new and is in many ways unavoidable. The practice of science, as all practices, depends on institutional funding, and thus will always face some threat of corruption by external goods. Nearly a century ago, Weber said that research institutes were state capitalist projects,[17] and Cold War genetics depended on military funding. Yet while we should not look back nostalgically to some golden age before *Diamond v. Chakrabarty*, there does seem to have been an intensification of this process. The production of scientific knowledge has become increasingly dependent on capital, so institutions that were at one point devoted to more or less pure counterpractices to commercial society are now deeply embroiled with commerce.[18] At the same time, advanced modern science has never been able to function without access to such resources. Thus, even if one could transform the practice of science beyond the entrepreneurial model, any new structures would introduce their own temptations and deformations of character because they would depend on some form of institutional support providing external goods.

An Aristotelian could accept all this and just expel the practice of biology, chemistry, or mechanistic physics from the virtuous community. Yet such a move would defeat some of the strengths of this position, because such traditions aim at developing a better understanding of humans and our world, which must include the understanding gained through science.[19] Thus, a model of moral formation must be developed that allows science to continue but forms the virtuous scientist in ways that counter problems in standard scientific formation and those that would arise under any future development of the structures of scientific research.

A solution might come from a better understanding of why an isolated community might be necessary for virtuous moral formation. This isolationism arises in part from an interpretation of Aristotelian moral formation. As contrasted with a common interpretation of Plato in which Plato supposedly holds that knowledge of the good is sufficient for virtue or

[17] Max Weber, "Science as a Vocation," in *From Max Weber*, ed. H. H. Gerth and C. Wright Mills (New York: Oxford University Press, 1958), 131.

[18] Paul Rabinow, *French DNA: Trouble in Purgatory* (Chicago: University of Chicago Press, 1999), 20.

[19] Alasdair MacIntyre, "The End of Education," *Commonweal*, October 20, 2006, 14.

Aristotle's misinterpretation of Socrates as saying the same thing,[20] Aristotle insists that habituation is necessary to form the passions in virtue if one's practical wisdom is to function properly.[21] The Aristotelian soul contains both the intellectual faculties of the soul and prerational faculties of the passions. The passions can be made obedient to reason but only by training them, shaping them through repeated performance of good actions. "The virtues we get by first exercising them."[22] One forms one's passions in the habits of virtue by consistently choosing actions in line with virtue. The conscious choice of virtuous action inclines the prerational passions toward similar actions and desires in the future, thus gradually transforming the affections toward virtue.

Since the passions incline us toward the goods that one chooses, how does a person initially form these good habits? More importantly for this project and for a Christianity devoted to conversion and sanctification, how can one transform vicious habits once they are present in the adult? At some points, Aristotle does suggest that the adult can begin to choose the actions corresponding to a virtue even if he does not have that virtue: the enkratic or continent man chooses the action corresponding to virtue even if he does not have that virtue, warring with his passions to do so.[23] He masters his soul through his reason, and though this stage does not involve fully virtuous action, since virtue requires properly oriented passions, it is a step in that direction.[24] Even the incontinent man seems capable of reform: "Those who are incontinent through habituation are more curable than those in whom incontinence is innate; for it is easier to change a habit than to change one's nature; even habit is hard to change just because it is like nature."[25] There are possibilities of moral reformation in Aristotle's ethics, but they are extremely difficult.

Still, Aristotle generally seems to think that one must be habituated in virtue from early youth: "It makes no small difference, then, whether we

[20] John Sellars, *The Art of Living: The Stoics on the Nature and Function of Philosophy* (Burlington, VT: Ashgate, 2003), 52. This interpretation of Plato seems wrong, since Plato has a different conception of knowledge from the modern one. One achieves knowledge only after ethical work transforms the subject and allows him to remember the world of forms. Thus, subjectivation is a preliminary step to grasping the truth, a truth that is then spiritually transforming.

[21] For a good discussion of this issue, see Julia Annas, *The Morality of Happiness* (New York: Oxford University Press, 1993), 54–5.

[22] Aristotle, "Nicomachean Ethics," in *Introduction to Aristotle*, ed. Richard McKeon, trans. W. D. Ross (New York: Modern Library, 1992), bk. 2.1.

[23] Aristotle, "Nicomachean Ethics," bk. 7.9.

[24] Virtuous passions seek the goods of virtue and thus take delight in virtuous action. See Aristotle, "Nicomachean Ethics," bk. 2.3.

[25] Aristotle, "Nicomachean Ethics," bk. 7.10.

form habits of one kind or of another from our very youth; it makes a very great difference, or rather all the difference."[26] Ideally, one is gradually formed in these virtues by the laws and customs of the good *polis*, which adjust the citizen's passions from an early age through the proper play of pleasure and pain as incentives. Of course, one eventually must come to understand why one acts the way one does and to depend on reason in action in order to be fully virtuous, but the habituation of the passions through good communal practices serves as an essential preparation for virtue. If one is not raised in a good *polis*, then one has little hope of achieving virtue: "It was open at the beginning not to become men of this kind, and so they are unjust and self-indulgent voluntarily; but now that they have become so it is not possible for them not to be so."[27] This pessimism makes sense because there is little in this model that would allow for the reformation of the vicious passions that result from a flawed education. Other commentators have similar concerns about transforming the problematic aspects of contemporary character. For example, Albert Barrera points to the difficulties of changing those who have been steeped in the market order for their entire lives, an especially salient problem when dealing with the issues of entrepreneurial science.[28] Drawing on Aristotelian ethics, MacIntyre needs a community that is wholly directed toward forming virtuous individuals.

There is a second more recent source for the rejection of the possibility of virtuous formation in wider society: Marx. Beginning his career as a Marxist and always deeply influenced by Marx, MacIntyre was disillusioned by Stalinism and sought in Marx for ways to argue against the bureaucratic state of the Soviet Union and the vanguard party of Leninism.[29] MacIntyre argues that a deeper engagement with Marx's Third Thesis on Feuerbach would have prevented the problems of later Marx and Marxism.[30] The third thesis on Feuerbach states:

> The materialist doctrine that men are products of circumstances and upbringing, and that, therefore, changed men are products of other

[26] Aristotle, "Nicomachean Ethics," bk. 2.1.

[27] Aristotle, "Nicomachean Ethics," bk. 3.5.

[28] Albino Barrera, *Market Complicity and Christian Ethics* (New York: Cambridge University Press, 2014), 238.

[29] For MacIntyre's writings on Marxism, see Alasdair MacIntyre, *Alasdair MacIntyre's Engagement with Marxism: Selected Writings 1953–1974* (Boston: Brill, 2008). For the influence of Marxism on his later thought, see Kelvin Knight, *Aristotelian Philosophy: Ethics and Politics from Aristotle to MacIntyre* (Cambridge: Polity, 2007).

[30] Alasdair MacIntyre, "The Theses on Feuerbach: A Road Not Taken," in *The MacIntyre Reader*, ed. Kelvin Knight (Notre Dame, IN: University of Notre Dame Press, 1998), 232.

> circumstances and changed upbringing, forgets that it is men who change circumstances and that it is essential to educate the educator himself. Hence, this doctrine necessarily arrives at dividing society into two parts, one of which is superior to society ... The coincidence of the changing of circumstances and of human activity can be conceived and rationally understood only as revolutionising practice.[31]

This thesis argues against the idea that intellectuals or politicians will be able to introduce changes to society based on their theoretical conclusions because they themselves are embedded in these practices of society.[32] Truth cannot lead to transformation. Change can occur only if individuals are first formed by virtuous practices. This stance is not a Burkean conservatism derived from Marxist theses because it assumes that the foundations of contemporary society are deeply flawed and that widespread change is needed. However, these changes can begin at an individual level only if these individuals have been formed through practices that are found in good communities. Thus, one must turn to marginal communities that retain such a network of practice.

Other practice theorists, like those discussed in the last chapter, have confronted this same problem. Pierre Bourdieu argues that sociology could transform society if the discipline itself were transformed according to his findings.[33] Bourdieu argues for a reflexive sociology, one that takes intellectual and theoretical findings from research and applies them to the institutional structure of sociology itself, making it less hierarchical, less exploitative, more collaborative. If this field were purified, it could serve as a base from which to transform the rest of society through policy recommendations.

MacIntyre admits the attraction of this thesis, but he sees it as too elitist, as denying the formative effects of the broader society, and as too restricted in its disciplinary focus. Bourdieu

> relates theory primarily to the practice of the scientific enquirer, that is, of the theorist. But Marx's point, I take it, was that no solution from within theory or even from within the practice of theorists is possible. It is only from the standpoint of social practice of a very different kind, one prior to both enquiry and theory, that a solution will be possible.[34]

31 Karl Marx, "Theses on Feuerbach," in *The Marx-Engels Reader*, ed. Robert Tucker (New York: Norton, 1972), 108.

32 MacIntyre, "The Theses on Feuerbach: A Road Not Taken," 231.

33 Pierre Bourdieu and Loïc J. D. Wacquant, *An Invitation to Reflexive Sociology* (Chicago: University of Chicago Press, 1992).

34 MacIntyre, "The Theses on Feuerbach: A Road Not Taken," 231.

MacIntyre thus continues his attack on the reliance on technocratic knowledge and expertise. An epistemic elite threatens to form an illegitimate locus of power. The development of the university into increasingly specialized disciplines only privileges the creation of such an epistemic elite, preventing the development of a well-educated population capable of making decisions on a wide variety of issues. Social decisions cannot be left to an elite from one field but rather must be engaged by people from all crafts and walks of life in order for all information to be brought to bear on the problem and for all goods to be respected. Specialized knowledge must lose the privileges given to it by the modern emphasis on technique, where its possession of facts allows it to dictate the possible courses of action to those who use their values to choose between the presented options. All citizens must possess the requisite knowledge for a healthier community.

Such populist, anti-expertise, anti-technocratic strains are now a dominant influence in political theology. A similar example of a populist religious politics is Jeffrey Stout's politics of bottom-up community organizing, which rejects the need for an epistemic or intellectual elite to guide political action.[35] While Stout convincingly shows that community organizing is very useful for local political issues such as constructing sewers or preventing discriminatory police searches, it is not clear that it will be helpful for highly technical disputes over science, such as arguments over autism and vaccines, climate change, or monetary policy, which require a highly educated and specialized elite to inform and guide public action. Domains of specialized expertise are isolated from the larger game of democratic reason-giving because of the apprenticeship that they require.[36] Society and its problems are tremendously complex and thus many of its critical functions are structured around systems of knowledge and expertise, systems that have brought much good. Any changes to social systems to procure justice, empowerment, or mere survival require the input of experts. While MacIntyre's isolated communities are an explicit rejection of the modern nation-state and the position of science within it, Stout's seeming embrace of the contemporary US polity under the form of organized groups of citizens is just as much a rejection of the reality of

[35] Jeffrey Stout, *Blessed Are the Organized: Grassroots Democracy in America* (Princeton, NJ: Princeton University Press, 2010).

[36] For a debate in the science studies literature over this statement, see Harry Collins and Robert Evans, "The Third Wave of Science Studies," *Social Studies of Science* 32, no. 2 (2002): 235–96; Sheila Jasanoff, "Breaking the Waves in Science Studies," *Social Studies of Science* 33, no. 3 (2003): 389–400; Brian Wynne, "Seasick on the Third Wave? Subverting the Hegemony of Propositionalism," *Social Studies of Science* 33, no. 3 (2003): 401–17.

contemporary society and the role of expert knowledge within it. Both imagine a form of society different from that of bureaucratic modernity, with MacIntyre merely recognizing that fact. Neither engages the necessity of expert power for basic human well-being.

Richard Rorty sees similar problems with Stout's position when he argues that traditional political struggles such as the labor movement required both bottom-up and top-down organizational structures to mediate knowledge, provide conceptual elaboration, and successfully publicize these causes.[37] Stout acknowledges Rorty's rejection of his position but does not respond to Rorty's arguments.[38] Of course, Rorty is referring to the necessity of an older style of intellectual, not the kind of scientific experts discussed here, but his arguments are just as applicable to the expert positioned in contemporary power structures. Instead of the calls emanating from theology for a bottom-up politics incapable of addressing many of our most pressing issues, theological ethics and politics need to engage contemporary expertise. Such an ethics and politics would be for the scientist as scientist, instead of merely the scientist as citizen or Christian. Indeed, it would be for the scientist as scientist, Christian, and citizen.

The Specific Intellectual and the Care of the Self

Foucault, on the other hand, appreciates the importance of the expert in contemporary society. From this appreciation, a new figure capable of changing relations of power arose in Foucault's late thought, the "specific intellectual." This figure is not the person who escapes the apparatus of power, who exists outside society, or who exemplifies the resistor as the artist, musician, criminal, etc. This figure is the one who inhabits the apparatus and thus is able to redeploy it.

The concept of the specific intellectual is formulated in opposition to Jean-Paul Sartre, who was the consummate universal intellectual claiming the ability to speak on any issue.[39] Such a person is no longer possible because today's problems are so fundamentally intertwined with intricate technical issues of biology, climate, economics, etc. that their solutions can be formulated only by those with a narrow expertise or by groups of

[37] Richard Rorty, *Achieving Our Country: Leftist Thought in Twentieth-Century America* (Cambridge, MA: Harvard University Press, 1998), 53.

[38] Stout, *Blessed Are the Organized*, 298n16.

[39] Michel Foucault, "Truth and Power," in *Power*, ed. James Faubion (New York: New Press, 2001), 126.

experts from various fields working together. Just as ordinary citizens are no longer able to address these problems, neither can elite, generalist public intellectuals. Social solutions and changes must come in part from people with narrow technical expertise like geneticists, economists, or climatologists working in concert with other social movements. They are "specific" because they have specific expertise, they are tied to specific offices and laboratories, and they engage in specific localized struggles. This intellectual does not have the universal prudence that MacIntyre seeks, but she has a technical expertise and a position in the power structure that allows her to act as an agent of resistance by countering dangerous deployments of power such as nuclear weapons or eugenics. She can work to change how people think about issues.[40] In the present situation, such resistance could be directed at the neoliberal biotechnology regime's system of ownership, reward, and ideals of the scientific entrepreneur.

Such a position is not anti-expertise. It is precisely because it acknowledges the interrelation of power and knowledge that it is able to accurately represent the constraints that scientists face as well as see the importance of scientists for productive political action. Such a vision counters the populist strand embedded in much of political theology.

Yet, as noted in previous chapters, the scientist is formed in the current dominant disciplinary paradigm, and her knowledge will be used in flawed contemporary power structures. How could someone be reformed into a specific intellectual who can serve as an agent of change and justice? There are two tasks that must be accomplished in this formation. First, the scientist must be morally formed so that she will defend the good of truth internal to science against its replacement by external goods such as money or power. Second, the scientist must be formed so that she ensures that the knowledge she produces is not used to reinforce destructive power such as that entailed by a reductionist understanding of the world. These two ends are intertwined, since forms of knowledge and power, truth and governmentality, are interrelated. The question is how to realize this kind of moral formation.

Kelvin Knight suggests that a turn to individual practices might prompt action.[41] If an important aspect of bureaucratic modernity is the degradation

[40] Foucault, "Truth and Power," 133.

[41] Kelvin Knight, "Revolutionary Aristotelianism," in *Virtue and Politics: Alasdair MacIntyre's Revolutionary Aristotelianism*, ed. Paul Blackledge and Kelvin Knight (Notre Dame, IN: University of Notre Dame Press, 2011), 32.

of individual practices through their redirection by institutions from internal goods like truth toward external goods like wealth, then the response to the problems of modernity might be the mobilization of practitioners to defend the integrity of their own practice. Thus, if science becomes directed toward money rather than truth because of legal, financial, and academic institutions, then the response by researchers and the scientific community should be to lobby for structural changes that would redirect the field back toward knowledge.

At first glance, his suggestion seems promising, but there are problems that it fails to address. It assumes a pure group of practitioners, whereas wider society and scientific training already forms researchers within the entrepreneurial ideal. Moreover, it does not engage with the problem of the deployment of scientific knowledge in broader society to buttress contemporary forms of power by reducing persons, bodies, and nature to material for use. Finally, it ignores how specific practices internal to a practice like science may form one in ways that support a reductionist rationality. The Heideggerian technological worldview is imbued by everyday scientific techniques, techniques necessary for contemporary research to advance toward truth. This worldview and these techniques need to be addressed before one can form virtuous practitioners. As Northcott notes in a different context, it is important to reactivate communities to reform social organizations, but that alone is not enough, since a change of heart is also necessary.[42] The question then becomes what sort of intervention can form practitioners into the type of people who defend the true goods internal to their practice against institutional pressure.

Reformulated in this manner, Tony Burns notes that Knight's suggestion of focusing on the virtues of the individual practitioner seems similar to Foucault's notion of care of the self because this care allows the individual to form herself in virtue despite the negative pressures of the surrounding milieu.[43] Care of the self is the neglected counterpart of the Delphic command to know thyself.[44] In Foucault's eyes, modern philosophy, by focusing solely on the knowledge of what the human is, what one's personal truth is, or who one's authentic self is, has neglected

[42] Michael S. Northcott, *The Environment and Christian Ethics* (New York: Cambridge University Press, 1996), 300–8.

[43] Tony Burns, "Revolutionary Aristotelianism? The Political Thought of Aristotle, Marx, and MacIntyre," in *Virtue and Politics: Alasdair MacIntyre's Revolutionary Aristotelianism*, ed. Paul Blackledge and Kelvin Knight, 52.

[44] Michel Foucault, *The Hermeneutics of the Subject: Lectures at the Collège de France, 1981–82*, trans. Burchell (New York: Palgrave Macmillan, 2005), 3.

the ancient imperative to care for the self, to work on the self, and to transform the self. It is this self-work that links politics and ethics. "There is no first or final point of resistance to political power other than in the relationship one has to oneself."[45]

This call to care for the self runs through ancient philosophy starting with Socrates. In the *Apology*, Socrates says that his mission has been to call his fellow citizens to care for themselves.[46] The clearest call for care for the self, for the soul, though, appears in the *Alcibiades I*, a text that served as the first Platonic text read in ancient Platonic education.[47] The *Alcibiades I* clearly shows the relation between care of the self and defective education.[48] In the dialogue, Socrates attempts to convince Alcibiades to take up the study of philosophy if he wants to lead, because he has not had the moral education of the leaders of Athens's chief rivals, the Spartans and Persians. Socrates invites him to transform his soul to gain the ability to govern himself and others.

This concern over defective moral education is crucial for why this concept is helpful for addressing the problems of a malformed subjectivity and and the problematic ways that practices can shape how one sees the world. This care aims to transform a deformed character into a virtuous one. The emphasis on defective education becomes even greater in Stoicism.[49] The Stoic ideal subject was the sage, who had a totally coherent system of knowledge in agreement with divine Providence. Since the human soul was actually a piece of the divine, living rationally in accordance with nature was the fulfillment of the human's true being. Yet the Stoics recognized very few sages; though wisdom is in accord with human

[45] Foucault, *The Hermeneutics of the Subject*, 252.

[46] Foucault, *The Hermeneutics of the Subject*, 5–7.

[47] Here I will not engage the debate over Foucault's reading of *auto to auto* in *Alcibiades I* as "self itself." For discussion, see Brad Inwood, *Reading Seneca: Stoic Philosophy at Rome* (Oxford: Clarendon Press, 2005), 322–52; Christopher Gill, *The Structured Self in Hellenistic and Roman Thought* (New York: Oxford University Press, 2006), 325–407. While Foucault's reading of this passage may be an overinterpretation, I do not think that his formulations mean that he is positing a new ontological formulation of the self, nor that his use of the term "subject" must necessarily imply the Cartesian subject, which would run counter to much of Foucault's work. Even given misreadings regarding the self, his conclusion that a focus on caring for oneself requires a reflexive relationship still stands. How could one work on one's soul in a way that would not in some way be reflexive? Nor does Foucault necessarily deny that ancient ethics had an idea of an objective core to the person related to reason. See also Gretchen Reydams-Schils, "Review of Reading Seneca: Stoic Philosophy at Rome by Brad Inwood," *Phoenix* 61, nos. 1–2 (April 1, 2007): 186–9.

[48] Foucault, *The Hermeneutics of the Subject*, 37.

[49] Foucault, *The Hermeneutics of the Subject*, 95.

nature, almost no one realizes that nature.[50] Stoics explained this seeming contradiction through an error theory that is fundamentally developmental. Stoics, especially Chrysippus, argued that people are misformed either by bad social influence or by the deceptive attractiveness of things in the world.[51] These two elements cause the false judgments that value the truly indifferent elements of the world more highly than reason and virtue rather than situate them in their proper place in the system of human values. Care of the self reforms these defective passions, impulses, and representations. Unlike Aristotelians, the Stoics can encourage people to undertake this reform without undergoing an early habituation of the passions because the Stoic soul is not divided into intellectual and prerational faculties. Passions are fundamentally wrong judgments.[52] By transforming one's judgments that result from one's initial impressions of the world, one can change one's passions. Thus, altering a malformed character is possible in Stoicism, even if it is very difficult. This care transforms the subject in the direction of wisdom.

This transformation occurs through a process of *askesis*, of training and exercise. It requires taking on a whole style of life, a regimen involving medicine, sexual practices, and domestic life. In the classical era, in Plato's portrayal of Socrates, for example, the moral aspect of this *askesis* seems to have involved merely doing virtuous action, and, in this sense, is largely congruent with Aristotle's habituation and MacIntyre's formation through practice in a community.[53] It is with the Hellenistic and Roman period that the care of the self develops into something far more technical, "a differentiation between the exercises themselves and the virtue . . . for which they were meant to serve as training: their procedures . . . tended to form a particular technique that was more complex than the mere rehearsal of the moral behavior they anticipated."[54] Specific spiritual exercises develop with the explicit goal of transforming the subject. These are not merely virtuous activities in everyday life, but intensive efforts at work on the self, its passions, and its representations. These practices involve both meditations that are intended to transform how

[50] A. A. Long and D. N. Sedley, *The Hellenistic Philosophers* (New York: Cambridge University Press, 1987), 381.

[51] Annas, *The Morality of Happiness*, 178; Thomas Benatouil, *Faire Usage: La Pratique du Stoicisme* (Paris: Vrin, 2006), 115–18.

[52] Margaret Graver, *Stoicism and Emotion* (Chicago: University of Chicago Press, 2007); Martha Nussbaum, *Upheavals of Thought* (New York: Cambridge University Press, 2001); Sorabji, *Emotion and Peace of Mind: From Stoic Agitation to Christian Temptation.*

[53] Michel Foucault, *The Use of Pleasure* (New York: Vintage Books, 1985), 74.

[54] Foucault, *The Use of Pleasure*, 77.

one perceives the world and ascetic practices that test how far one's subjectivity has been transformed.[55]

These techniques of the self were taken up by early Christianity as seen in Clement of Alexandria, in Augustine,[56] and especially in the growth of monastic practice. They have continued into the modern period in the form of various spiritualities, such as Ignatian spirituality, private devotional prayer, fasting, and abstinence. In *Hope and Christian Ethics*, David Elliot discusses how the *ars moriendi* tradition of devotional exercises that served as a preparation for death used such spiritual exercises to form one in hope.[57] By focusing one's attention on the inevitability of death and on the struggles that one must go through, it drew the person's attention away from attempting to find ultimate satisfaction through merely worldly goods. The *ars moriendi* forced one to contemplate one's ultimate aim to achieve the beatific vision through practices like wandering through cemeteries to reinforce knowledge of mortality. As this example shows, Christian techniques of care of the self attempt to form kinds of subjects very different from Stoicism, but the formal theoretical structure remains similar.

It is the emphasis on these private devotional practices that makes Stoic ethics so amenable to adoption by Christian ethics. Much of the more communitarian thrust of recent discussions of virtue has little explicit place for these exercises that are essential to many forms of Christian life. While historically the private may have played too prominent a role in comparison to the communal in Christian ethics, perhaps the private has played too little of a role in recent work. Foucault's analysis allows the integration of this work on the self into understandings of the virtuous Christian life, which will of course always occur in the context of the Church. This self-work allows the layperson to prepare herself to witness to Christ in pluralist environments and to critically engage with other formative influences.

55 For a thorough discussion of these practices and their use in shaping character, see Pierre Hadot, *Philosophy as a Way of Life: Spiritual Exercises from Socrates to Foucault*, trans. Arnold I. Davidson (Cambridge: Blackwell, 1995); Ilsetraut Hadot, *Sénèque: Direction spirituelle et pratique de la philosophie* (Paris: Vrin, 2014).

56 Sarah Catherine Byers, *Perception, Sensibility, and Moral Motivation in Augustine: A Stoic-Platonic Synthesis* (New York: Cambridge University Press, 2013), 151–71.

57 David Elliot, *Hope and Christian Ethics* (New York: Cambridge University Press, 2017), 154–9. For other recent discussions of the *ars moriendi* in Christian ethics, see Christopher P. Vogt, *Patience, Compassion, Hope, and the Christian Art of Dying Well* (Lanham, MD: Rowman & Littlefield, 2004); Allen Verhey, *The Christian Art of Dying: Learning from Jesus* (Grand Rapids, MI: Eerdmans, 2011); Lydia Dugdale, ed., *Dying in the Twenty-First Century: Toward a New Ethical Framework for the Art of Dying Well* (Cambridge, MA: MIT Press, 2016).

Ethics, Community, and Truth in Stoicism

However, many theologians strongly object to this model of moral formation. It seems too Kantian, too individualist, too praxis-oriented, too self-involved. Foucault said little about how he envisioned these techniques acting in contemporary life, and the little he did say was directed toward the practices of pleasure. Even aside from questions of sexual norms, given the centrality of the Cross for the Christian subject, pleasure is not a likely ethical focus for Christian ethics.[58] Moreover, his complex concepts of truth and the subject have strong divergences from Christian ideas of these topics, though these differences are less prominent than they may first appear. Because of these concerns, here I follow in Foucault's tracks to focus on Stoic care of the self to defend these ideas, although I will return to some of these problems in Chapter 7. Stoicism itself, however, has been challenged on at least three fronts.[59] Perhaps the best way to demonstrate the utility of the concept of care of the self to contemporary Christian ethics is by responding to the primary criticisms of Stoic ethics.[60] These revolve around Hellenistic engagements with rules,

[58] That said, Christians must also be careful before too quickly rejecting his work on techniques of the self merely because he mentions sadomasochism and must not reduce his work to an embrace of power and death as in Jim Miller, *The Passion of Michel Foucault* (New York: Simon & Schuster, 1993). David Halperin argues that such a move disqualifies his thought based largely on his sexuality in David Halperin, *Saint Foucault: Towards a Gay Hagiography* (New York: Oxford University Press, 1995).

[59] My response to these critiques draws on A. A. Long, "Greek Ethics after MacIntyre and the Stoic Community of Reason," *Ancient Philosophy* 3 (1983): 184–97; Annas, *The Morality of Happiness*; David Burchell, "Civic Personae: MacIntyre, Cicero and Moral Personality," *History of Political Thought* 19, no. 1 (1998): 101–11.

[60] Another method of proceeding would be to show that Stoic philosophical principles strongly influenced early Christianity, especially the New Testament, since Stoicism was the predominant popular philosophy of the period. For examples of this approach, see Troels Engberg-Pedersen, *Paul and the Stoics* (Louisville, KY: Westminster John Knox Press, 2000); Troels Engberg-Pedersen, *Cosmology and Self in the Apostle Paul: The Material Spirit* (Oxford: Oxford University Press, 2010); Tuomas Rasimus, Troels Engberg-Pedersen, and Ismo Dunderberg, *Stoicism in Early Christianity* (Peabody, MA: Hendrickson Publishers, 2010); Runar Thorsteinsson, *Roman Christianity and Roman Stoicism: A Comparative Study of Ancient Morality* (Oxford: Oxford University Press, 2010). I do not engage this strand of appropriation in part because it has faced strong criticisms in the field of biblical studies, such as in C. Kavin Rowe, *One True Life: The Stoics and Early Christians as Rival Traditions* (New Haven, CT: Yale University Press, 2016). For other explorations of Stoicism in Christian ethics that differ in their interpretation of Stoicism, see John Bowlin, *Contingency and Fortune in Aquinas's Ethics* (New York: Cambridge University Press, 2010); Elizabeth Agnew Cochran, "Consent, Conversion, and Moral Formation: Stoic Elements in Jonathan Edwards's Ethics," *Journal of Religious Ethics* 39, no. 4 (2011): 623–50; Elizabeth Agnew Cochran, "Bricolage and the Purity of Traditions: Engaging the Stoics for Contemporary Christian Ethics," *Journal of Religious Ethics* 40, no. 4 (2012): 720–9; Elizabeth Agnew Cochran, *Protestant Virtue and Stoic Ethics* (London: Bloomsbury, 2017). For an engagement with Foucault's

community, and truth. This section will show that the Stoics were not the forerunners of an overly simplified Kantianism, that their individual development was intimately connected to the care of and interaction with the other, and that work on the self is a work of truth.

Stoic Rules

A first critique of the Stoics leveled by an entire tradition that rejects casuistry is that their ethics is an ethics of rules and law rather than virtue, with the multiple virtues of Aristotle reduced to the single virtue of following rules. This critique sees the Stoics as precursors to Kant: "Indeed whenever the virtues begin to lose their central place, Stoic patterns of thought and action at once reappear. Stoicism remains one of the permanent moral possibilities within the cultures of the West."[61] For these critics, rules separated from teleology and virtue will neither transform character nor bring coherence to a life. This critique evinces a misunderstanding of the role of rules in Stoic ethics.

Stoics themselves argued over the role played by specific precepts versus character and general philosophical principles in ethics.[62] Seneca relates this debate in terms of a distinction between *decreta*, the basic philosophical principles of Stoicism, and *praecepta*, instructions for specific situations in life.[63] Aristo of Chios, an early Stoic who was a disciple of Zeno, argued that one did not need *praecepta*, but merely the *decreta* and virtue. Specific rules will always need modification and interpretation, so it is better just to follow general principles.[64] Against this position, Chrysippus argued that, of course, *praecepta* needed to be justified and interpreted by *decreta* and that the sage may be able to govern herself only by *decreta*, given her deep

practices of the self that maintains a strong communitarian emphasis, see Jonathan Tran, *Foucault and Theology* (New York: Bloomsbury T&T Clark, 2011).

61 MacIntyre, *After Virtue*, 170.

62 Seneca, *Epistles 93–124*, revised edition (Cambridge, MA: Loeb Classical Library, 1925), Ep. 94, 95; Annas, *The Morality of Happiness*, 94–106; Inwood, *Reading Seneca*, 95–131.

63 Seneca, *Epistles*, 94.1–2. For a different interpretation of *praecepta* as admonitions to select rather than to pursue preferred indifferents and to recognize differences in situations rather than in specific instructions, see Katja Maria Vogt, *Law, Reason, and the Cosmic City: Political Philosophy in the Early Stoa* (New York: Oxford University Press, 2008), 189–213.

64 The sage would be justified in performing all kinds of seemingly immoral actions like cannibalism and incest depending on the situation, such as eating dead or amputated human flesh if it were the only way to survive, or committing incest if a father and daughter were the only humans remaining after a global catastrophe. These examples recall medieval discussions of the sins of the Patriarchs. For a discussion of these "disturbing theses" of Stoicism, see Benatouil, *Faire Usage*, 302–20; Vogt, *Law, Reason, and the Cosmic City*, 20–64.

understanding of Providence. For those who had not yet achieved wisdom, however, *praecepta* are extremely helpful for understanding how *decreta* are translated into daily life. Rules of conduct become the material for translating philosophical principles into an *ethos*. Stoics were not governed by general laws, since Providence is specific, nor by mere rules, since rules of conduct were to be the mode of transformation of character that could always be further interpreted. These rules function very much like rules in Aristotelian virtue ethics.

Part of the problem is that many of these critics draw their interpretation of Stoicism only from Cicero, who translated the Stoic term *katechon* as *officium*, which is often translated as duty with its Kantian overtones. *Katechon* cannot mean Kantian duty, since plants and animals also have *katechonta*.[65] These are behaviors that conform to nature and are thus close to a teleological vision. Applying *katechonta* is like using an Aristotelian syllogism.[66] As with other norms, one must rearrange them as one learns about conflicts between norms.

The critique of Stoic rules can advance in another form, under the idea that these rules are rather conventional. "In no fundamental way . . . do the Stoic accounts of law and justice provide any grounds for an appeal against the dominant practices or order of Greek and Roman politics," since the Stoics limited duties to others to those obliged through reciprocity.[67] If one looks merely at Stoic ideals, then this criticism is unfair. It in part results from reading Cicero as a completely accurate representation of Stoic thought, rather than seeing Cicero as attempting to bring Stoic ethics more in line with the *mos maiorum*, traditional Roman ethics.[68] If one considers true Stoics, their ethical guidance could be quite unconventional with revolutionary potential. They bequeathed to the Western tradition the concept of the equal dignity of every human being based in a shared rationality. The emperors certainly saw some threat in these Stoics since they exiled Seneca, Musonius Rufus, and Epictetus and forced Seneca and Thrasea Paetus to commit suicide. Yet Stoics were committed to the idea that common human conceptions of the good and duty were not

[65] Benatouil, *Faire Usage*, 121.
[66] Benatouil, *Faire Usage*, 124.
[67] MacIntyre, *Whose Justice? Which Rationality?*, 149.
[68] Marcia L. Colish, *The Stoic Tradition from Antiquity to the Early Middle Ages* (New York: E. J. Brill, 1990), 61–158; Gretchen Reydams-Schils, *The Roman Stoics: Self, Responsibility, and Affection* (Chicago: University of Chicago Press, 2005), 94. As Paul Veyne says, "For Cicero, philosophy was simply a mouthpiece for convictions and attitudes acquired elsewhere." See Paul Veyne, *Seneca: The Life of a Stoic* (New York: Routledge, 2002), 15.

completely wrong because nature planted the seeds of virtue into human reason.[69] Any philosophy that accepts this idea cannot help but keep some conventions, however revisionary it might be. Moreover, any philosophy will be trapped in part by the intellectual milieu in which it arises.[70]

Community

This critique of the conventionality of Stoic politics is connected to a second major strand of criticism, a criticism of the seeming turn toward the individual subject and away from the community. A common critique sees the distinctiveness of Hellenistic ethics as arising from its unique political culture, arguing that Stoicism arose at the same time as the great Hellenistic kingdoms, which displaced the *polis* and removed the *polis*'s elite from the activity of governing.[71] Faced with a loss of freedom and threats from these larger structures, the elites withdrew into themselves. Philosophy changed from the classical *polis*-centered thought of Plato and Aristotle into a focus on the subject, reflecting this new individualism. Given this loss of control over circumstances and the dangers of fate in the new kingdoms, the elites sought happiness not in activity but in tranquility and detachment from others. Thus, there was a loss of community, politics, and interpersonal relation, echoing changes seen today. With the loss of participatory small communities in the face of technocratic, managerial bureaucracies, people withdraw from the community into individualism. Instead of seeking happiness in virtue and right interpersonal relations, people seek it through consumption disguised as an aesthetics or realization of the self. The cult of the self is a result of and support to contemporary power structures in both antiquity and today, resulting in impotent, self-absorbed subjects.

This picture of Hellenistic culture is incorrect and thus leads to a faulty picture of Stoicism. I will address this picture on three levels: the relation between the government of self and others, the dependence on the other for self-formation, and the difference between the Stoic turn inward and

69 Annas, *The Morality of Happiness*, 107; Martha Nussbaum, *The Therapy of Desire: Theory and Practice in Hellenistic Ethics* (Princeton, NJ: Princeton University Press, 1996), 332; Benatouil, *Faire Usage*, 236.

70 In his study of Stoicism and Christianity in Rome in the first century, Thorsteinsson argues that early Christians were just as unrevolutionary in regard to institutions like slavery, and may have been less universalizing than the Stoics. See Thorsteinsson, *Roman Christianity and Roman Stoicism.*

71 Alasdair MacIntyre, *A Short History of Ethics: A History of Moral Philosophy from the Homeric Age to the Twentieth Century*, 2nd edition (Notre Dame, IN: University of Notre Dame Press, 1998), 99–108; MacIntyre, *After Virtue*, 170.

the contemporary cult of the self. To address the first level, current scholarship suggests that there was no simple loss of power by the elite in the *polis*. With the rise of the great empires, the *polis* lost its independence, but not its importance as a locus of government since the kingdoms and empires operated through the institutions of the *polis*.[72] The local elites retained a great deal of power.

This structure made the ethics of the local elites no longer the ethics of citizens who had full power over themselves and struggled for power in the *polis*. Instead, the elites occupied intermediate positions, always with the powers of officials above them and power over officials or locals below them. It was an ethics for intermediaries.[73] Here, Foucault and MacIntyre's different visions of the contemporary situation shape their visions of the past. Faced with the imperial edifice of power, the elites became impotent for MacIntyre, much as the modern citizen is impotent in the face of the nation-state and its expertise-driven bureaucracy. For Foucault, the elites' embedding in an apparatus of power made for a complex ethics involving the negotiation of different forms of power and subordination, much like that of the modern specific intellectual. It provides a lens for seeing contemporary ethics under a different form.

In this model, the government of the self is always tied to the government of others. In the *Alcibiades I*, the spur to self-care and self-knowledge is Alcibiades's ambition to rule the city. The situation is different once the subject of power becomes an intermediary. There is always a balance that must be struck between the relations to superiors and to inferiors. For example, Seneca's *De beneficiis* is a long treatise on giving and receiving benefits in a system of patron–client relations. These concerns are thus best placed in a social setting of hierarchy. Because of this social setting, Hellenistic philosophy may provide good models for our current age. Instead of offering the vision of complete entrepreneurial freedom that neoliberalism provides, it sees individuals enmeshed in hierarchies of power that they constantly negotiate. The scientist as specific intellectual is always beholden to institutional and funding powers, be they universities, corporations, or grant review committees. Yet the scientist also operates in positions of authority over the careers of students, technicians, and those colleagues whose papers and grants she reviews.

72 See the later, late Antique picture of the relationship between urban elites and empire in Peter Brown, *Power and Persuasion in Late Antiquity: Towards a Christian Empire* (Madison: University of Wisconsin Press, 1992).

73 Foucault, *The Care of the Self*, 88.

If modernity is bureaucratic, then Hellenistic ethics offers a good starting point for our reflections.

It does this by distancing one from any single social role that one assumes. All such roles are hostages of fortune instead of fundamental aspects of the self. They can be lost. "Take care that you are not turned into a Caesar," writes Emperor Marcus Aurelius.[74] Being the emperor is merely a job, a vocation, like any other with its own special duties and rewards. Instead of alienating one from these roles, this distancing is meant to help one perform a role better. If one associates oneself too closely with such a role, then one is tempted to abuse the power of that role and to be crushed when it is lost. This distancing is thus not merely a rejection of social roles. Instead, as Gretchen Reydams-Schils has argued, the Stoic self was a mediator between different roles and the value systems tied to these roles.[75] The Stoic understood the self as embedded in systems of relationships, all of which had their own duties attached to them. One is a father, a senator, a Roman, a human, and, most importantly, a rational being. Practices of the self helped the individual to construct a self that could fulfill the demands of all his roles, not as abstract duties but as a way of living a harmonious, happy life.

This Stoic concept of the self is complex, and one can achieve a proper conception of the self as it relates to its roles, to its needs, and to others only through a developmental process called *oikeiosis*, or appropriation. The self does not merely include the individual human in her own body, but also includes all those connected to her in some way.[76] The paradigmatic experience of this connection is in parenthood, where one's child is seen as deeply, intimately connected to the parent's self.[77] The self expands outward to one's city or nation. Because one is a rational being, one eventually should appropriate all other rational beings as connected to

[74] Marcus Aurelius, *Meditations*, trans. Robin Hard (New York: Oxford University Press, 2011), bk. 6.30; Reydams-Schils, *The Roman Stoics*, 1.

[75] Reydams-Schils, *The Roman Stoics*, 2.

[76] Here, I do not read *oikeiosis* as a dual process of personal and social *oikeiosis* as in Annas, *The Morality of Happiness*, 265. In this reading, there are two separate processes that the scholar or Stoic must reconcile just as moderns attempt to reconcile egotism and altruism. Instead, I follow Reydams-Schils and others as reading the Stoic self as deeply, ontologically embedded, and, thus, personal and social appropriation are two aspects of fully coming to know the self. The reading of them as divergent processes came from their description in separate passages of *De Finibus*, but nowhere does Cicero explicitly say that they are two separate processes that must be harmonized.

[77] Marcus Tullius Cicero, *On Moral Ends* (New York: Cambridge University Press, 2001), bk. 3.62–3; Gisella Striker, "The Role of Oikeiosis in Stoic Ethics," *Oxford Studies in Ancient Philosophy* 1 (1983): 145–67; Mary Whitlock Blundell, "Parental Nature and Stoic Oikeiosis," *Ancient Philosophy* 10 (1990): 221–42.

oneself. This connection is not just a glorification of the abstract rational capacity. In Stoic physics, it is an ontological connection, with every person existing as an actual piece of the divine. Despite the oddness of the ontology, this conception is not as alien to Christianity as it might seem. This view is similar to seeing the image of God ontologically embedded in all rational creatures who can consciously know and love themselves.[78] In Augustinian Christianity as well, it is this reflection of the divine in our free and rational capacity that can impart dignity and connection.

Yet this appropriation of oneself and others as rational is not easy. While the prerational appropriation of animals and children to their bodies and immediate loved ones occurs fairly automatically, rational *oikeiosis* takes effort. It takes a lifetime of "pains and study" (*cura*) to acquire virtue.[79] It is relatively easy to see how one learns to care for one's family, but it is difficult to see how one cares for distant citizens and foreigners. Hierocles the Stoic develops a response to this problem. He envisions the self and its relations as a series of concentric circles.[80] At the center is the individual and her body, the next circle out is her parents, then immediate relatives, then distant relatives, extending to other citizens and even foreigners. At first glance, this vision is similar to Aquinas's order of charity,[81] and for Hierocles, this picture does reflect one's natural affections and understanding of duty. However, Hierocles's point is that this picture must be actively transformed by drawing the concentric circles ever closer to the self, or, rather, by expanding one's conception of the self to include more circles.

One does this through intentional practices of the self. It is done gradually, "with effort," one circle at a time, rather than in an immediate conversion leading to the recognition of universal brotherhood. Of course, this includes acting with love toward others, but it also includes exercises of naming and altering representations. One begins to call more distant family members by more intimate names – referring to aunts as mothers or to cousins as siblings.[82] Such a practice acts upon our conceptions, strengthening social relations. This process continues until the most distant person is included in one's vision of the self. Thus, these Stoic duties

78 Thomas Aquinas, *Summa Theologica*, trans. Dominicans of the English Province (New York: Benziger Bros., 1947), I.93.

79 Seneca, *Epistles 93–124*, 124.14.

80 Hierocles, *Hierocles the Stoic: Elements of Ethics, Fragments and Excerpts*, trans. Ilaria Ramelli (Atlanta, GA: Society of Biblical Literature, 2009), 91.

81 Aquinas, *Summa Theologica*, II–II.31.3.

82 Hierocles, *Hierocles the Stoic: Elements of Ethics, Fragments and Excerpts*, 93.

are realized through the transformation of affections and the understanding of the self, not a cold understanding of the moral law. It is a logic of identification through human connection to the rational divine nature. Whatever one might think of the efficacy of these particular practices for the intended outcomes, an efficacy of which I am doubtful, it does provide a presentation of the complex model of Stoic *oikeiosis* and opens the way for other techniques that might achieve the same goal. One acts upon the self through spiritual practices to achieve a transformation of character.

Because of defective human education, this transformation requires not only work but also another person to initiate and guide this work. Care of the self is an interpersonal practice that always requires a spiritual guide.[83] The guide knocks one out of one's routine and helps one to progress. One might take up the care of the self by attending a school of philosophy. However, Stoic schools, unlike the Platonic Academy, the Aristotelian Lyceum, or the Epicurean garden in which the student could spend his entire life, were schools that individuals attended as training or as a retreat before returning to their duties in the wider world.[84] Very few of the students aimed to become professional philosophers or to retreat from the world. The schools trained one for autonomy. Stoic schools thus have far more points of contact with contemporary educational institutions than other ancient philosophical institutions. These techniques of the self are never free floating, but always take place within some tradition of thought that gives structure and definition to the relations among truth, power, and the self.

Finally, despite the relational nature of the care of the self in its Stoic guise, it still results in a turn inward. The spiritual guide ultimately aims to make the disciple autonomous, to allow her to care for herself.[85] The disciple gains self-mastery, delights in the self, and sees the self become the subject of truth. Just because it focuses on a turn inward should not necessarily keep these techniques from the Christian repertoire. Trinitarian reflection in the Augustinian tradition is posited on such an inward turn toward the image of God in ourselves that points us toward the actuality of God. Gilles Emery has described the meditative relation between the concepts of the Trinity and the *imago dei* in Augustine and Aquinas as a spiritual exercise:

[83] For a thorough study of how such spiritual guidance worked in practice, see Hadot, *Sénèque*.

[84] Gretchen Reydams-Schils, "Authority and Agency in Stoicism," *Greek, Roman, and Byzantine Studies* 51, no. 2 (May 28, 2011): 296–322.

[85] Reydams-Schils, *The Roman Stoics*, 18.

> Augustine emphasizes ... that in order to glimpse God, the spirit must purify itself of corporeal representations ... This is precisely the usefulness of the study of creatures, and the goal of the exercise. The *exercitatio* proposed by Augustine is an *ascension* toward God from the image that is inferior and unequal to him, and it is at the same time a gradual movement *toward the interior.*[86]

It is a work on the self in which a deeper understanding of one's being and its reflection of the divine lead to the transformation of the image within to greater perfection through an awakening of love for God. Thus, Stoic care of the self is a possible model for Christians in bureaucratic modernity. It is a turn inward that allows one to conform more closely to a divine ideal that then allows one to use the power of a position in an apparatus for service to others and to respond correctly to the powers that impinge on the self. It is a way of always taking the other into account as the self is transformed.

Truth

The third and most important critique of the care of the self for this project is its seeming dismissal of truth. To many, this concept seems to devalue *logos*, contemplation, and rationality in favor of *ethos*, practice, and will. MacIntyre laments that "among Socrates' disciples there were some ... who stylized the Socratic way of life and drew their moral code from this style of life rather than from reflection on the character of definition."[87] What was distinctive about Socrates was that he challenged Athenians' definitions of virtue, arguing that they did not know what virtue is and thus could not achieve virtue. By focusing on Socrates' style of life, Hellenistic ethics threatens to continue this ignorance of virtue. The concern is that there is no tether in reality to the proposed self-transformation. Into what is one to be transformed? There is a danger that

[86] Gilles Emery, *Trinity, Church, and the Human Person: Thomistic Essays* (Naples, FL: Sapientia Press, 2007), 46.

[87] MacIntyre, *A Short History of Ethics*, 101. Martha Nussbaum sees this decreased attentiveness to reason as a problem not with the original Stoic sources but with their interpretation by Hadot and Foucault; Nussbaum, *The Therapy of Desire*, 5. This concern is also found in Maria Antonaccio, "Contemporary Forms of Askesis and the Return of Spiritual Exercises," *Annual of the Society of Christian Ethics* 18 (1998): 69–92. John Sellars notes that this is a problem with Hadot's work but not with Foucault's analysis, which is deeply concerned with the interplay of ethics and truth; Sellars, *The Art of Living*, 116–17. A similar argument over the rationality of these practices is found in Daniel del Nido, "Pierre Hadot on Habit, Reason, and Spiritual Exercises," *Journal of Religious Ethics* 46, no. 1 (March 1, 2018): 7–36; John Cooper, "Socrates and Philosophy as a Way of Life," in *Maieusis: Essays in Honor of Myles Burnyeat*, ed. Dominic Scott (New York: Oxford University Press, 2007), 20–42.

truth based in practice will just reduce to performance.[88] At worst, care of the self, as in the form of Foucault's aesthetics of existence, seems to point to the most extreme form of self-will, a complete control over one's own end, which would be the triumph of the technocratic mindset.[89] Such a result would make Foucault's work and Stoic ethics worthless for Christian ethics since it would exalt human pride.

This diagnosis is true neither for Stoic ethics nor for Foucault's ethics. Stoicism saw their logic, physics, and ethics as organically interrelated.[90] To understand the dignity of human rationality, one must grasp the ontological link between the hegemonic part of the human soul and the divine. Further, Stoic *apatheia*, the absence of passions understood as irrational judgment, is achieved through correct judgment, in part by understanding the inevitability of certain events and trusting in the goodness of Providence. Similarly, the whole point of Foucault's later work was to show that power, specifically government, the forming of the subject by another or oneself, is always tied to a truth.[91] Care of the self *requires* some kind of truth. Foucault's focus on this care rather than the corresponding truth is due to his desire to recover this neglected part of philosophy, but he always recognizes that care for thyself and knowing thyself are linked in ancient philosophy.

Indeed, the care of the self seems so alien to questions of truth only because we have denied a strong relationship between the form of our subjectivity and truth. Foucault's genealogy of the relation between the subject and truth put us in the aftermath of what he calls the Cartesian moment.[92] By this, he does not merely mean the publication of Descartes'

[88] Fergusson, *Community, Liberalism and Christian Ethics*, 70–1.

[89] Brian Brock, *Christian Ethics in a Technological Age* (Grand Rapids, MI: William Eerdmans, 2010), 145. See also Arnold I. Davidson, "Ethics as Ascetics: Foucault, the History of Ethics, and Ancient Thought," in *The Cambridge Companion to Foucault*, ed. Gary Gutting (New York: Cambridge University Press, 1994); Hadot, *Philosophy as a Way of Life*, 206–14.

[90] Here I follow A. A. Long, *Hellenistic Philosophy; Stoics, Epicureans, Sceptics* (New York: Scribner, 1974); Sellars, *The Art of Living*, 80; Reydams-Schils, *The Roman Stoics*; John Cooper, "Eudaimonism and the Appeal to Nature in the Morality of Happiness: Comments on Julia Annas, *The Morality of Happiness*," *Philosophy and Phenomenological Research* 55, no. 3 (1997): 587–98; rather than Annas, *The Morality of Happiness*, 80, or Gill, *The Structured Self in Hellenistic and Roman Thought*, 78, who distance Stoic ethics from Stoic physics and metaphysics.

[91] Michel Foucault, *Du Gouvernement des Vivants: Cours au Collège de France (1979–1980)*, ed. François Ewald, Alessandro Fontana, and Michel Senellart (Paris: Seuil, 2012), 6.

[92] Foucault, *The Hermeneutics of the Subject*, 14. Sellars, Nehamas, and others bring out what is implicit in Foucault, that while the ancient versus modern split is a convenient shorthand, it oversimplifies the history of philosophy; Sellars, *The Art of Living*, 2–3; Alexander Nehamas, *The Art of Living: Socratic Reflections from Plato to Foucault* (Berkeley: University of California Press, 1998). Aristotle separated theory and practice, intellectual and moral virtue and thus had little need for the transformation of the subject to attain truth. He had no spirituality as Foucault says in

Meditations, but a whole historical trajectory tracing the separation of theology from spirituality, philosophy from theology, and the development of the scientific method. This trajectory stretched from the twelfth to the nineteenth century, with its defining event in Descartes,[93] but its initial wedge in scholastic theology. This trajectory resulted in a view that truth is achieved through an allegedly neutral method. One does not need to be a certain kind of person to achieve truth, but merely must apply the correct techniques to nature, history, society, etc.[94] Truth has no costs for the subject.

The truth of ancient philosophy, on the other hand, was costly. "For the subject to have right of access to the truth he must be changed, transformed, shifted, and become ... other than himself. The truth is only given to the subject at a price that brings the subject's being into play."[95] Truth demanded that the subject be transformed by the extensive *askesis* of the care for the self. The subject must efface the bad education that deforms his capacity for truth and, through difficult exercise, reconstitute himself in a form capable of achieving this truth. Once obtained, truth will in turn transform the subject, thus providing happiness.[96] These different understandings of truth contradict each other: ancient truth gave access to Being, which transforms one, while Cartesian truth is knowledge of objects in the domain of things, a reductionist vision of truth.[97]

Stoic *askesis* made one the subject of truth, solidifying one's grasp of a complex system, leading to coherence, virtue, calmness, and the ability to appropriate oneself as a member of the rational community of gods and men. Stoic techniques of the self attempt to translate philosophical truths into systematic behaviors. These techniques form the acting subject like a

The Hermeneutics of the Subject, 17. The relation between these aspects of *eudaimonia* in Aristotle are of course a matter of contention. Gabriel Richardson Lear in particular argues that they should be more closely tied than they are in contemporary Aristotelian scholarship, saying that practical virtue was a way of participating in contemplation throughout one's daily life. See Gabriel Lear, *Happy Lives and the Highest Good: An Essay on Aristotle's Nicomachean Ethics* (Princeton, NJ: Princeton University, 2004). Yet Sellars points out that Aristotle misinterprets Socrates as saying that pure propositional knowledge brings about virtue, whereas Socrates meant that knowledge in the form of an art of living embedded in the subject through *askesis* will bring virtue. See Sellars, *The Art of Living*, 52. This misinterpretation supports the idea that Aristotle lacked a relation between conscious *askesis* and truth, relying instead on habituation.

93 Foucault, *The Hermeneutics of the Subject*, 25–7.

94 Foucault, *The Hermeneutics of the Subject*, 18, 190; Michel Foucault, *The Government of Self and Others* (New York: Palgrave Macmillan, 2010), 20.

95 Foucault, *The Hermeneutics of the Subject*, 15.

96 Foucault, *The Hermeneutics of the Subject*, 16.

97 Foucault, *The Hermeneutics of the Subject*, 191. Here one can see the influence of Heidegger.

hammer and chisel used to shape a statue, transforming *logos* into *ethos*.[98] It takes the philosophical knowledge learned through study and actually makes it effective in one's life by making it ready to hand, available for use in the moment of action.

For the Stoics, these *logoi* were largely the *decreta* and *praecepta* discussed earlier, but Reydams-Schils has discussed the importance of a second form of these truths, that of natural knowledge.[99] This form focused on knowledge as ethopoetic, knowledge that could actually give rise to an *ethos*.[100] Seneca gives an example of meditation on such doctrines in his *Natural Questions*.[101] Here, Seneca criticizes many types of knowledge, seeing history as completely useless except for a few exemplars of virtue, such as Cato. The good knowledge, the knowledge that one must investigate, is that knowledge which pertains to divine Providence and its plan for the interrelationship of all things in nature.

> Once you have ascended to those truly great regions, whenever you see armies marching with standards ... it is a mere scurrying of ants toiling in narrow confines ... Up above there are vast spaces, which the mind is allowed to enter and occupy, provided that it takes scarcely anything of the body with it, that it wipes away any uncleanness, and that it soars upward unencumbered, nimble, and self-reliant. When it has reached those regions, it finds nourishment, it grows, and, as though freed from its chains, it returns to its origin. It has this proof of its own divinity, that it takes delight in the divine and enjoys it not as someone else's possession but as its own.[102]

By taking a God's-eye view, by seeing the order and rationality of the cosmos, one gains the knowledge that will put the world and one's life into perspective. First, one sees oneself caught in an ordered causal nexus. Next, by looking at events from the point of view of the universe, one sees the ultimate meaninglessness of battles, wealth, and all the things normally valued. Finally, one grasps that the rationality displayed in the universe is not alien to humanity but rather is actually the same as human rationality. In other words, it helps one to appropriate the universe.

Appropriation does not occur merely by possessing knowledge. "Clutch with both hands ... make them a part of himself, and by daily meditation

[98] Foucault, *The Hermeneutics of the Subject*, 327.

[99] Reydams-Schils, *The Roman Stoics*, 40.

[100] Foucault, *The Hermeneutics of the Subject*, 237.

[101] For the ethics in Seneca's *Natural Questions*, see Gareth D. Williams, *The Cosmic Viewpoint: A Study of Seneca's Natural Questions* (New York: Oxford University Press, 2012).

[102] Lucius Annaeus Seneca, *Natural Questions*, trans. Harry M. Hine (Chicago: University of Chicago Press, 2010), bk. 1.10–3.

upon them to bring himself into such a state of mind, that these wholesome maxims occur to him of their own accord, that ... they may straightway be ready for use when required."[103] As seen in the prior description, appropriation occurs when one receives knowledge and, at the same time, meditates on the meaning of that knowledge. One visualizes how oneself and one's surroundings fit into place in the universe, and one sees both the insignificance and the grandeur of one's life. One must actively appropriate this ethopoetic knowledge through meditation, turning it from philosophical or natural knowledge into grounds for action in the world.

Such activity also bears on perception. Marcus Aurelius recommends taking every object that enters into one's vision and pausing on it, meditating on it.[104] One contemplates its cause, its connections to the cosmos, its uses, its value to oneself. In so doing, one inserts the object into a total system of representation in a way that will hopefully lead one day to the harmonizing of the entire system. At that point, one will see nature harmoniously, and thus one may act with nature harmoniously, and attain virtue. Here again, the Stoic is taking the knowledge of an object and actualizing it, using an exercise to make it real. He is transforming it into a path toward virtue.

Stoics also examined themselves. At the beginning of the day, they recommended trying to foretell the challenges and activities of the day in order to equip oneself to meet them effectively and virtuously.[105] At the end of the day, one was to look back over the day and take account of one's activities to see how one progressed or failed over the day.[106] These exercises prepared one for action. Preparation for the future also included *praemeditatio malorum*, the anticipation of evils.[107] One placed oneself meditatively in the position of suffering, both to prepare oneself to meet that eventuality and to take stock of the true value of things in one's life. Going through the exercises of properly envisioning the correct place of things in the cosmos is essential to living correctly. It turns dead knowledge into living dispositions.

[103] Seneca, *Seneca: Moral Essays*, vol. III: *De Beneficiis*, trans. John W. Basore (Cambridge, MA: Loeb Classical Library, 1935), bk. 7.2.1.

[104] Foucault, *The Hermeneutics of the Subject*, 290–8.

[105] Marcus Aurelius, *Meditations*, bk. 1.15.

[106] Epictetus, *Discourses, Books 3–4. Fragments, The Encheiridion*, trans. W. A. Oldfather (Cambridge, MA: Harvard University Press, 1928), bk. 3.10.2–3; Seneca, "De Ira," in *Moral Essays*, vol. I, trans. John W. Basore (Cambridge, MA: Harvard University Press, 1928), 3.36.

[107] Cicero, *Tusculan Disputations*, trans. J. E. King (Cambridge, MA: Harvard University Press, 1927), bk. 3.14.

Stoic care of the self allows one to become a subject of truth who consciously embodies a philosophical system. Here, it is not unconscious habituation that ensures right action but the recalling of truth. These truths are not just cognitive but have become embodied through meditation and *askesis*. The Stoic constitutes herself as a rational actor, thus realizing herself as part of the divine, living harmoniously and with tranquility.

Christianity shifts the form of this relation since the transformation of the subject by truth is the first step.[108] Ethical training does come before baptism, but in general, it is the gift of grace that is based in the truth received in faith that transforms the subject, enabling her to act on herself in charity. Yet one's grasp of truth in charity is further deepened through both works of mercy and devotional spiritual exercises. This growth in faith is accompanied by a renunciation of the old man of the flesh in order to instantiate oneself in the new man of Christ. As such, in many Christian traditions, it requires intensive work of the self on the self.

The relationship between the manner of *askesis*, or style of life, and a philosophy or system of truth is never simply deductive:

> I would like to emphasize that this relationship between metaphysics of the soul and stylistics of existence is never a necessary or unique relationship ... The relations between the two are flexible and variable... There is a certain degree of constancy in the metaphysics of the soul specific to Christianity, you know very well that Christianity developed very different styles of existence ... within the framework of this metaphysics ... The ascetic's life is not everyone's life; the lay person's life is not the same as that of the cleric; the life of the monk ... is not the same as that of the secular clergy, and so on.[109]

Systems of truth can engender many different styles of life capable of attaining them. This trait gives Foucault's ethics a formal character, one that tends to be nonnormative, which is actually quite helpful. One can use nonnormative formal concepts to understand other traditions, such as Judaism, Islam, or various philosophies. From an ecumenical perspective, it allows helpful discussions of the differences between Reformed subjectivity and Lutheran subjectivity, for example. It allows one to see multiple styles of life and forms of subjectivity even within the Catholic tradition. A Thomist subjectivity, a Franciscan subjectivity, and a charismatic subjectivity will all be different and differently formed, none adjusting itself

108 Foucault, *The Hermeneutics of the Subject*, 255.

109 Michel Foucault, *The Courage of Truth: The Government of Self and Others II: Lectures at the Collège de France 1983–1984*, trans. Graham Burchell (New York: Palgrave Macmillan, 2011), 164.

to their shared moral norms and dogmatic beliefs in the same way. None of them will be wrong for that reason, and all of them will stand in partial contrast to the dominant entrepreneurial subject. While subjectivity follows truth, there can be a diversity of subjects within a broadly shared conception of truth.

This analysis shows the fertility of the care of the self for thinking about the subject's relation to truth in philosophy and religion. The real question is whether it can have any bearing on the attainment of scientific truth. After all, science is the epitome of a subject-independent method. My argument in the last part of this chapter, an argument prefigured in previous chapters, is that scientific truth is also costly for the subject, requiring the subject's transformation. From this conclusion, one can ask about the relationship between the subjective transformations required by scientific and religious truth and how Christian techniques of the self can be brought to bear to ease the combination of these two subjects of truth in a single person.

The Care of the Self in Science

This analysis of care of the self may seem adequate for theological and perhaps even philosophical endeavors, but one may doubt that it has implications for other disciplines in the academy. Is not the apprenticeship in the sciences or humanities enough of a moral formation? While a basic apprenticeship may be enough to practice an intellectual discipline as a career, there is another strand of reflection on the intellectual life that demands more. Erwin Chargaff commented that most practitioners of science are not true scientists; they are merely specialists.[110] The true scientist is denoted by the intensity with which he pursues truth and mystery. These reflections point to exemplary figures who devote themselves to the vocation toward truth in their field. This line of thought believes that science as a vocation requires adopting a different style of life and extensive self-work. In these discussions, one can see precursors to the call in this study to use practices of the self to transform the character of the entrepreneurial scientist.

The classic Catholic example of such an analysis is Antonin-Gilbert Sertillanges's *The Intellectual Life*. While not specifically addressing natural science, he believed that the pursuit of truth requires adopting a particular

[110] Erwin Chargaff, *Heraclitean Fire: Sketches from a Life before Nature* (New York: Rockefeller University Press, 1978), 114.

regimen of life. The intellectual must seek silence.[111] She must purge her life of unnecessary things and social engagements,[112] pursuing detachment so that her mind might dwell in simple clarity.[113] In Sertillanges, the entire day is structured according to certain practices of the self that will help to advance the scholar's pursuit of truth. On waking, she anticipates the day's tasks: "The morning is sacred for thought."[114] When she tires, she exercises in nature to restore herself for more work and to keep her body in the proper condition for intellectual work.[115] While walking, she looks for truth in everything she sees, analyzing her representations in a very Stoic manner.[116] In the evening, she puts the work in order for the next day before relaxing in quiet conversation.[117] As she falls asleep, she plants the germ of a problem in her mind so that the mind might work during the unconscious hours of the night.[118] Sertillanges recommends a diet and regimen that mortifies the body in order to purify the senses and thus the mind. The vocation to truth requires another kind of life, with a training through *askesis*. Here, too, attaining truth is costly, but Sertillanges frames it in terms of salvation coming only through the Cross.[119]

This ascetic vision appears even in secular writings, such as Weber's "Science as a Vocation." While his most famous asceticism regards the rejection of meaning, the devotion to facts to the exclusion of values, any such devotion requires more practical forms of *askesis*. One must accept humiliation and exclusion from the prevailing fashion of the discipline and from one's academic institution.[120] For a long time, perhaps forever, the scholar lives in poverty, a poverty allowing freedom to work.[121] The scholar should do his own calculations, engaging in the hard labor of the discipline, even if he is senior enough that it is not necessary.[122] Moreover, the intellectual devotes himself completely to his question. "Whoever lacks the capacity . . . to come up to the idea that the fate of his soul depends upon whether he makes the correct conjecture at this passage of the

[111] A. G. Sertillanges, *The Intellectual Life: Its Spirit, Conditions, Methods* (Westminster, MD: Newman Press, 1959), 43.
[112] Sertillanges, *The Intellectual Life*, 39.
[113] Sertillanges, *The Intellectual Life*, 37.
[114] Sertillanges, *The Intellectual Life*, 69.
[115] Sertillanges, *The Intellectual Life*, 35.
[116] Sertillanges, *The Intellectual Life*, 60.
[117] Sertillanges, *The Intellectual Life*, 72.
[118] Sertillanges, *The Intellectual Life*, 68.
[119] Sertillanges, *The Intellectual Life*, 174.
[120] Weber, "Science as a Vocation," 134.
[121] Weber, "Science as a Vocation," 131.
[122] Weber, "Science as a Vocation," 135.

manuscript may as well stay away from science."[123] Yet the scholar also knows that his answer will be superseded and forgotten.[124] Only *askesis* prepares one for the critical interrogation of one's own and others' work.

This scientific *askesis* has a literary memorial in Sinclair Lewis's *Arrowsmith*, a scientific bildungsroman.[125] While there are many fictional portrayals of the scientific life, *Arrowsmith* has a relevance for contemporary scientists that others lack. It was given to me by my undergraduate thesis advisor, the person who first trained me in research, as a graduation gift to show his preferred model of the scientific vocation. It follows the maturation of Martin Arrowsmith from medical school to his mature scientific life. During his undergraduate studies, he is inspired by the exemplary hero Max Gottlieb working late in the night on a project that involved imparting relentless criticism of his own and others' work. Gottlieb seeks a mathematical, analytical understanding of life and disease. Such a heartless understanding of the world is, to him, the only way to achieve truth, and truth is the only way to ultimately improve human life.

The book is structured as a series of temptations that Arrowsmith must overcome to fulfill his vocation to truth. The temptations of bodily pleasures like alcohol must be overcome to enable a focus on work. Marriage to a wealthy woman threatens to submerge him in the obligations of high social status. The mercantile respectability of common medical practice requires overbilling, over-operating, and cultivating appearance over reality. Service in government health departments requires blunting one's actions to achieve popularity and political success. Even scientific success leads to the temptation to publish too rapidly in order to capture fame and funding, and it involves the threat of promotion to positions of administrative responsibility that keep one from research. Wealth, social position, and family life create responsibilities that keep one from the laboratory. Finally, the temptation that Arrowsmith does not overcome is that of compassion in the face of an epidemic, compassion that leads him to neglect including a proper control group in a study of a prophylactic against plague. This compassionate action means that the experiment is inconclusive and that the utility of the prophylactic is unknown.

In this novel, the scientific life is pictured as a series of struggles and renunciations. It is eulogized by Gottlieb:

[123] Weber, "Science as a Vocation," 135.
[124] Weber, "Science as a Vocation," 138.
[125] Sinclair Lewis, *Arrowsmith* (New York: Harcourt Brace Jovanovich, 1952).

> To be a scientist – it is not just a different job ... It is a tangle of very obscure emotions, like mysticism ...; it makes its victim all different from the good normal man ... The scientist is intensely religious – he is so religious that he will not accept quarter-truths, because they are an insult to his faith. He want that everything should be subject to inexorable laws ... He must be heartless. He lives in a cold, clear light. Yet dis is a funny t'ing: really, in private, he is not cold nor heartless – so much less cold than the Professional Optimists. The world has always been ruled by the Philanthropists ... and see once what a fine mess of hell they haf made of the world! Maybe now it is a time for the scientist, who works and searches and never goes around howling how he loves everybody.[126]

Arrowsmith overcomes these temptations through *askesis*. The novel does not have the same kind of explicit meditative techniques of the self discussed earlier, but it still portrays a work of self on self in order to become a subject of truth. He renounces career after career, comfort after comfort. He devotes himself to mastering mathematics and interpreting the world through it. He constantly examines himself. In the end, he breaks with the standard scientific career path, leaving the New York institute of which he could be chair in order to take up a life of research with a few close collaborators in the Vermont woods. There, in a manner surprisingly similar to Sertillanges's vision of the work of thought, scientific research is mixed with exercise in nature and physical labor. Pursuing even truth as secular reductionism is costly.

Despite many similarities in ascetic practices, there is a fundamental difference between Sertillanges's Christian vision of the scientific life and the secular vision of Weber or Lewis. At first glance, all seem to encourage a flight from the world into contemplation. While all three see scientific truth as useful to the common good, with Weber advising policy makers, Arrowsmith curing disease, and Sertillanges's truth serving the public, Weber and Lewis seem to encourage the scientist to cut all ties, to become a sort of monk. Actual human relations become things that one must see through. Weber's devotion to truth ends in clarity and freedom from illusions of meaning.[127] It is a vision of the world as a network of mathematical causation.[128] From Lewis's descriptions, the scientist is set apart, even alienated, from others. Research animals are brutally used. Human experimental subjects are engaged coldly and endangered so that the experiment will work. It is an ascetic transformation to a subject of

126 Lewis, *Arrowsmith*, 266–7.
127 Weber, "Science as a Vocation," 151.
128 Weber, "Science as a Vocation," 139.

mechanistic truth without meaning. In its form of *askesis*, it is more a Platonic divorce from the world rather than a Stoic model of social embedding.

In contrast, Sertillanges's scholar, though seeking purity, silence, and nondistraction, must also be embedded in the world. "You will not imagine that . . . an increase of intellectual possibilities should prevail over the achievement of your true self . . . Do what you ought and must; if your human perfection requires it, the different demands it makes will find their own balance."[129] The active life of political engagement develops total virtue.[130] Social duties serve as inspiration, and Sertillanges even addresses marriage and children.[131] He embeds the scientist in a network of social duties in an attempt to develop a way to integrate the self as scholar, spouse, parent, and citizen.

Sertillanges's scholar also engages in a second array of techniques of the self that seek to embed scholarly truths within the greater Christian truth. The scholar's work should constantly be analyzed to fit into this broader framework of truth because the scholar is a subject of revealed truth whose vocation lies in drawing out truth from nature. This integration is furthered by prayer and devotional practices:

> The ideal would be to establish in one's mind a common life of thoughts all interconnected and forming as it were only one. So it is in God . . . The spirit of contemplation and prayer that we have pleaded for would of itself bring us nearer to that state . . . By taking God's point of view thanks to which everything finds its ultimate dependence and all things their cohesion, one cannot but feel oneself at the center of things.[132]

The day begins and ends with devotional meditation and prayer. The practice of finding truths everywhere through examining one's representations seeks not only mechanistic truth but moral and theological truth in every experience. Work and rest take the form of prayer.

Wisely, Sertillanges does not describe or prescribe a particular form of devotion, meditation, or spirituality. The scientific life can be combined with the great diversity of Christian spiritualities and styles of life:

[129] Sertillanges, *The Intellectual Life*, 49.

[130] Sertillanges, *The Intellectual Life*, 52–4.

[131] Sertillanges, *The Intellectual Life*, 40–1. Unfortunately, Sertillanges restricts himself to a conservative view of a woman's possibilities in scholarship. Women serve only to support their husbands' vocation. Children are more of a hindrance than a well-integrated part of the intellectual life. There is little that I would salvage from his explicit recommendations in these areas, but he is more cognizant than many other authors of the difficulties of balancing research and family life.

[132] Sertillanges, *The Intellectual Life*, 103.

Reformed, Methodist, Lutheran, Dominican, Franciscan, Ignatian, feminist, or ecological spiritualities. There will be certain commonalities: each will use practices to attempt to capture a more holistic understanding of the created order rather than a reductionist understanding, and each will value truth independently of material rewards. Yet each will lead to different shapes of the regimen of life and modes of practices. Each would serve to integrate one's work into a style of life that one freely takes up, transforming one's work and one's way of being in the world.

Though dependent on tradition and others, the realization of this style of life will be inescapably an individual affair dependent on self-work and taking a unique form for each. While this conception can be integrated into something like MacIntyre's community of practice or Hauerwas's church, it will be different because the scientist belongs inseparably to two different communities. The church is not her only home in the world. Though the church provides the integrating framework and ultimate truth for this style of life, the scientist spends most of her time enacting practices in institutions that are not formed by Christian principles and that are frequently at odds with those principles. In this way, she lives a Christianity adjusted to her particular station, drawing on aspects of individual Christian practice to fight the adverse effects of her daily practice by forming her perception. It is her scientific education and colleagues that provide her usual community, which makes the scientist's imitation of Christ quite distinctive.

The Christian is not fully at home in science either. Realizing possible problems with formation and the frequent hostility of his colleagues to religion, or more commonly their estimation of religion as an anachronism, the Christian knows that he must equip himself to deal with the daily struggles to retain a Christian vision of life and charity toward others. Though taking scientific ideals and heroes as his own, while embracing the ascetic scientific quest for truth, the Christian knows that these ideals, heroes, and quests must be modified by a vision of Creation and transcendent truth to be fully integrated into an imitation of Christ.

This style of life as individual struggle may be the inevitable form of the lay Christian engaging in a secular vocation in a non-Christian society. It is the life of difficult pilgrimage through the world. Through this stylization, she creates something new and beautiful. To Christians, she is an example of Christian devotion to Creation and truth in all its forms, drawing the best of secular knowledge into the Church. To her scientific colleagues, she serves as an example of Christianity's commitment to truth in all its forms and of the beauty of the Christian way of life. Yet this style can arise only from the personal struggle over the draw of two different forms of life, two different kinds of community, and the creation of a new way of imitating Christ from that struggle.

Conclusion

Scientists must take up the task of forming themselves in line with a scientific ideal if they want to confront the problems of contemporary science. These problems are linked to a moral exemplar, the entrepreneurial ideal, so resistance to them and, ultimately, their solution depends on realizing a different ideal in the self. The kinds of individual spiritual exercises discussed here will not be sufficient on their own to transform science. As previous chapters argued, much depends on the interrelationship between social structures and individual character. Transforming the research culture will require many institutional and policy changes of the kind discussed in Chapter 2. Yet these changes will not be successful, and may even be detrimental, unless they aim at a new kind of moral ideal whose realization will require the kinds of practices discussed here. Moreover, proper structural change can be set in motion only by those specific intellectuals who have already shaped their character using these kinds of practices. Therefore, a successful reform of scientific practice will depend in multiple ways on individuals embracing a different form of life. Until such reforms are initiated, though, individual scientists will need to undertake these kinds of spiritual exercises in order to act as good scientists in the face of the pressures of entrepreneurial science.

The integrations of science and religion discussed in the last section of this chapter should interest the scientist in regard to this new scientific ideal because the lack of this integration threatens to deform the scientific life, for two reasons. First, as has been noted, in the absence of a greater framework of meaning, mechanistic reductionism opens the way for the world to be treated as a mere instrument for use, an instrument easily adopted by power in the form of eugenics, weapons, etc. The decomposition of the world into discrete elements that can be deployed at will also opens the way for its commodification. If nature is not seen as a teleologically ordered system or a set of teleologically ordered creatures, then it can be exploited by neoliberal capitalism. The most important point of resistance is the relation of the self to the self, one's representations, affections, desires, and actions. It is by forming herself into an ethical subject that the scientist as specific intellectual can use her knowledge and position of expert power not to reinforce existing power relations but to change them.

Second, as Weber powerfully argued in *The Protestant Ethic and the Spirit of Capitalism*, this-worldly ascetic practice pairs quite readily with the entrepreneurial *ethos*. Self-denial and radical devotion to a task are good ways to make money. Since scientific findings are now commodifiable, the devoted scientist easily becomes the paradigm of the neoliberal entrepreneur. The ultimate goal of the late night in the laboratory pursuing the next result can shift from truth, to a patentable technique, and then to the success of one's start-up. Scientists should be concerned about the easy convergence between scientific practices and neoliberal ideals. Once the external good of money becomes the prime motivation for the growth of knowledge, the truth of research is threatened. Chapter 6 turns to the virtue arising from a true style of life that is necessary to combat these dangers.

CHAPTER 6

Acquiring the Virtue of Truth-Speaking in Science

As Chapter 5 discussed, the aim of the Stoic care of the self is to embody and live a system of truths. Part of living the truth is speaking the truth. One might think that speaking scientific truth unlike other kinds of truth has few moral or political stakes, given the Weberian understanding of the neutrality of science. In fact, speaking scientific truth is a central site of contemporary political conflict. The dangers to speaking scientific truth arise from two sources. While Chapters 1 and 2 discussed the first of these sources, the incentive structures internal to science that contribute to fraudulent or careless research, there are also pressures external to the scientific research enterprise that attempt to shape what kind of research results scientists publish. This chapter will discuss how larger interest groups deform the scientific research used for public decision-making to the detriment of the common good. Corporate and political interests influence what information is released about matters of public interest, such as new drugs or carcinogenic chemicals, in order to protect their products. Beyond merely blocking knowledge from public attention, such interests attempt to generate doubt about results that are released, as has been seen in the cases of tobacco and climate change. Scientists who oppose these interests are attacked on political, professional, and legal grounds. It now requires great courage to speak the truth on scientific matters concerning the common good.

Studies of philosophy in Antiquity describe how the quality of *parrhesia*, the courage to speak the truth in situations of danger, was an integral part of philosophical life, especially as it related to the public role of the philosopher.[1] This quality was necessary to challenge the person's own

[1] Peter Brown, *Power and Persuasion in Late Antiquity: Towards a Christian Empire* (Madison: University of Wisconsin Press, 1992); Michel Foucault, *The Government of Self and Others* (New York: Palgrave Macmillan, 2010); Michel Foucault, *The Courage of Truth: The Government of Self and Others II: Lectures at the Collège de France 1983–1984*, trans. Graham Burchell (New York: Palgrave Macmillan, 2011).

temptation toward laziness and falsehood, but it is demonstrated especially by the philosopher who contested the actions, beliefs, and forms of life of the politically powerful. Like most virtues, this courage of truth does not just arise naturally, but is formed by the care of the self described in the last chapter. Through this care, one becomes so tied to truth that one must proclaim it whatever the risk. This *parrhesia* was not just a quality of pagan philosophers but was taken up and modified by the Jewish and Christian traditions. In Christianity, it kept its sense of boldly proclaiming truth despite risk, becoming a characteristic feature of the apostolic witness to the gospel, while adding the sense of confidence and trust in God. These two aspects of proclamation and trust are connected, so the courage of truth takes on added importance for the Christian scientist.

This chapter will use *parrhesia* as an example of how the care of the self forms qualities essential to the contemporary scientist. It will begin by describing the contemporary problems of agnotology, the attempt by external actors to generate doubt about scientific research that threatens their interests. It will then characterize the role of *parrhesia* in the ancient philosophical tradition before turning to its use in Judaism and Christianity. Finally, it will discuss how this concept might be applied to the contemporary researcher, the kinds of exercises one must do, and the challenges one must face in speaking the truth in the contemporary world.

Creating Doubt

The response of the agrochemical corporation Syngenta to the findings of the Berkeley scientist Tyrone Hayes about its pesticide atrazine demonstrates the dangers that threaten scientists performing public interest research.[2] Internal Syngenta documents brought to light in a lawsuit over contaminated drinking water show that, after Hayes identified potential dangers of atrazine, Syngenta's public relations team attempted to discredit Hayes and his research. To this end, they performed a psychological evaluation. They coached or paid seemingly neutral third parties, such as professors, op-ed writers, and think tanks, to present their case to the media. When Hayes was considered for a job at Duke University, a Syngenta vice-president contacted one of the deans. Syngenta sent representatives to his talks to ask hostile questions. It even considered

[2] Dashka Slater, "The Frog of War," *Mother Jones*, February 2012; Clare Howard, "Pest Control: Syngenta's Secret Campaign to Discredit Atrazine's Critics," *100 Reporters* (blog), June 17, 2013, http://100r.org/2013/06/pest-control-syngentas-secret-campaign-to-discredit-atrazines-critics/; Rachel Aviv, "A Valuable Reputation," *The New Yorker*, February 10, 2014.

investigating Hayes's wife, and Hayes alleges that a Syngenta agent made veiled threats against his family.[3] The attacks took a toll on Hayes, in terms of both his psychological well-being and his career, but he continued to publicize his findings.

Hayes did not engage in this dispute out of hostility to agribusiness. His initial findings on atrazine arose through a research contract with Syngenta, an example in which a potential conflict of interest clearly did not affect the research findings. Once his data seemed to show that atrazine disrupted frog sexual development, Syngenta researchers began to attack his findings and delay funding for further studies necessary for publication. Hayes repeated the studies on his own and published them independently because of his commitment to truth. "Science is a principle and a process of seeking truth. Truth cannot be purchased and, thus, truth cannot be altered by money. Professorship is not a career, but rather a life's pursuit."[4] Hayes pursued truth as part of his vocation to science. It is this commitment to speaking the truth that Hellenistic philosophers denoted as *parrhesia*. *Parrhesia* was not a minor virtue because, as discussed below, it was tied to the form of the philosophical life, which was devoted to living the truth and, as with science, pursuing the truth. Moreover, as with Hayes, *parrhesia* involved speaking the truth in the face of risk.[5] Structural conditions have made truth-speaking in public interest science increasingly risky.

Syngenta's response to Hayes reflects a common corporate strategy. Its goal was not to disprove Hayes's findings and replace them with a more accurate account, an aim that would be perfectly legitimate given disputes in the scientific literature over the effects of atrazine. Instead, Syngenta sought to create doubt about his results by discrediting him personally. Robert Proctor coined the term "agnotology" to name the study of such active production of ignorance by social actors through the generation of doubt about scientific results.[6]

[3] Hayes responded with emails to Syngenta officials containing disparaging rap lyrics. Syngenta complained to UC Berkeley, but an investigation found no ethics violation. See Howard, "Pest Control," and Slater, "The Frog of War."

[4] Aviv, "A Valuable Reputation."

[5] Such cases are distinct from whistleblowing because of their tie to the vocation of science. Whereas whistleblowing reveals unethical practices within an organization, these conflicts concern different interpretations of reality, and the researchers involved frequently have no or indirect ties to the corporation. Many people attacking environmental research believe it is false. See Naomi Oreskes and Erik Conway, *Merchants of Doubt* (New York: Bloomsbury Press, 2010).

[6] Robert Proctor and Londa L Schiebinger, *Agnotology: The Making and Unmaking of Ignorance* (Stanford, CA: Stanford University Press, 2008).

The paradigmatic case of the production of doubt was provided by the tobacco industry, which even created a handbook for their strategy.[7] In 1953, facing evidence that cigarettes caused cancer, tobacco industry executives met with the public relations firm Hill and Knowlton.[8] This firm concluded that, while it would be impossible for the industry to disprove the mounting evidence of cigarettes' harm, the industry could create an atmosphere of doubt in the public mind that would hamper legal and regulatory action. Because of the centrality of this strategy to its continued survival, one industry executive wrote that "doubt is our product."[9]

How does one generate doubt? Positively, industries can create alternative centers of scientific research, such as the Tobacco Industry Research Committee in 1954, which became the Council for Tobacco Research in 1964.[10] These centers used grants to amplify the voices of sympathetic scientists and those pointing to other causes of lung cancer. Funding cultivated friendly researchers who would criticize negative research and investigators as part of an aggressive campaign to disseminate the message of doubt. Since no research is perfect, such analyses merely needed to show that a study is not absolutely conclusive and that other questions can be raised. These tactics created the appearance of scientific uncertainty on the question of whether cigarettes cause lung cancer, a conclusion that was untrue, but which stymied regulatory action. Companies also engaged friendly think tanks and grassroots organizations to publicize this seeming doubt. In response to research on secondhand smoke's ties to cancer, the tobacco company Philip Morris joined with other companies to form the Advancement of Sound Science Coalition in 1993, a seeming grassroots organization that opposed the use of environmental research in government regulation.[11] Think tanks aim at popular and policy audiences by lobbying Congress, writing books, and providing experts to appear on television.

The goal of this strategy is not to disprove mainstream science but to create a sense that there is doubt and debate about the field, and that there

[7] Oreskes and Conway, *Merchants of Doubt*, 144.

[8] This account of the tobacco strategy draws on Oreskes and Conway, *Merchants of Doubt*, 10–35, 136–68.

[9] USCF, Truth Tobacco Industry Documents, Brown & Williamson Records, Tobacco Industry Influence in Public Policy, Minnesota Documents, "Smoking and Health Proposal," 1969, http://legacy.library.ucsf.edu/tid/rgy93f00.

[10] Robert Proctor, *Golden Holocaust* (Berkeley: University of California Press, 2012), chap. 16.

[11] For details, see Oreskes and Conway, *Merchants of Doubt*, 150–2.

is conflict that does not actually exist in the scientific community. This goal is aided by the media's attempt to show both sides of an issue, even when the issue is scientifically settled, as in the case of anthropogenic climate change and the lack of connection between vaccines and autism. The battle also takes place in scientific journals through letters to the editor over specific articles. The general aim of this strategy is to deconstruct the scientific evidence by making complex criticisms of research methods and theoretical frameworks, even though these methods and frameworks may be widely accepted in the field. This strategy differs from standard scientific criticisms in that it attacks parts of experiments that cannot be tested and does not suggest experiments that would address their criticisms.

These tactics undermine the public consensus necessary for costly regulation. Instead of science presenting facts for politicians to analyze in light of their values, it appears that science just generates confusion. In general, these maneuvers can only delay regulation, because in the long run, research becomes too strong to ignore. For example, state governments targeted smoking through bans in enclosed workplaces starting in 1995, increased taxation, and a legal settlement in 1998.[12] But in the delay before such action, delays that often last decades, industries reap billions of dollars of profit, lives are lost, the environment is degraded, and the common good suffers.

This strategy to create doubt shows that even some of the tools meant to detect fraud, sloppy science, or conflict of interest can have unintended consequences because they can also be used deceptively to create doubt. One popular policy response to problems in science in an age when "information wants to be free" is to encourage more openness: research and trial data should be accessible by the public and the scientific community. Advocates of openness usually neglect the special nature of shared scientific training and the trust in the community that it produces in the analysis of scientific evidence. There is no one scientific method that can distinguish sound science from junk science. Every subdiscipline and every historical era has its own notion of proper method. Moreover, even within agreed-on methods, no experiment is perfect, and no scientific theory can be proven absolutely true. Much relies on the fact that data are analyzed and papers are reviewed by a community that shares the same training,

[12] California enacted the first workplace smoking ban with AB 13 in 1995. The 1998 Master Settlement Agreement between tobacco companies and forty-six attorneys general can be found at http://publichealthlawcenter.org/topics/tobacco-control/tobacco-control-litigation/master-settlement-agreement.

paradigms, and sense of common purpose. Without the induction into that community and its traditions of understanding this craft, it is hard to tell valid from invalid science.

Two legal enactments that give greater access to scientific data have aided the deconstruction of science. The Shelby Amendment of 1999 gave the public access to all federally funded research on which regulatory enactments are based, which gives industry groups access to all of the scientific data and notebooks for experiments that seem to threaten their interests in order to search for problems with the research. Providing this material is also extremely onerous to the scientist. The Information (or Data) Quality Act, which was hidden in an appropriations bill in 2001, allows anyone to "file a petition demanding 'correction' of information 'disseminated' by an agency, including scientific studies, if that person believes the information is not reliable, not objective, lacks utility, or is biased."[13] This language has been interpreted to mean that any study used to enact regulations must be perfect, which no scientific study is. Industries use these rules to attack climate change assessments, environmental regulations, and dietary guidelines. Thus, businesses can use the freedom of information given by these enactments to deconstruct scientific research, cast doubt on the work of researchers, and stop action in the face of this uncertainty.[14] Openness can lead to an increase in ignorance.

The potential payoff of this strategy of doubt can be seen in the way scientific research is used in court. Given the difficulties of establishing regulations and the lack of premarket or postmarket testing of chemicals in the United States, the legal system is crucial for environmental and product safety. Lawsuits remove products from the market both through the threat of compensatory and punitive damage awards and by bringing problems to the attention of regulators. They also serve commutative justice by providing funds for medical treatments or toxic cleanup. With the growing attention to environmental threats in the 1970s and 1980s

[13] Thomas O. McGarity and Wendy E. Wagner, *Bending Science* (Cambridge, MA: Harvard University Press, 2008), 151.

[14] This problem led Sheila Jasanoff to reverse her earlier emphasis on openness of research in public decision-making found in works such as Sheila Jasanoff, "Breaking the Waves in Science Studies," *Social Studies of Science* 33, no. 3 (2003): 389–400. "At some point . . . valid claims reach a state of repose that *should* preclude further questioning of the way they were produced . . . When public health and safety are at stake, such needless production of uncertainty could be . . . downright dangerous" (Sheila Jasanoff, "Transparency in Public Science: Purposes, Reasons, Limits," *Law and Contemporary Problems* 68 [2006]: 36).

came an explosion of liability lawsuits, so-called toxic torts, whose large settlements could bankrupt whole firms, as seen in the asbestos industry.[15]

To confront this threat, industry-funded think tanks mounted an attack on the research used to show that chemicals and products caused the specific injuries claimed in the lawsuits. The science used by plaintiffs was tarred as "junk science," a category that, according to these advocates, stood apart from the sound science of the regular scientific community. Junk science is supposedly produced by experts to serve the needs of plaintiffs' attorneys and is thus compromised by conflicts of interest.[16] These arguments are not entirely wrong. First, there are broad concerns about the accuracy of scientific research in general as discussed in Chapter 1. Specific problems emerge when one looks at scientific testimony in courts, since many questionable experts have been used in trials, and some companies have paid settlements for products that were eventually proven safe. Finally, even well-done public interest science is noticeably distinct from general scientific research. Alvin Weinberg called such public interest science "trans-science" since it can give important insights and guidance, but not the finality claimed by other fields because well-controlled experiments that would prove a connection between chemicals and cancer in humans, for example, are unethical.[17] This observation does not disqualify this science as bad science; it merely means that our methods have limited ability to attain certainty in these areas. Yet we have no other option than to depend on such research to guide policy.

Scientific testimony has generated similar debates since it was first introduced in court.[18] Expert scientific testimony is based on constantly developing theories and methods, leading to greater disagreement among experts than is otherwise the case. As far back as the nineteenth century, distinguished scientists on opposite sides of patent and insurance cases presented contradictory evidence, undermining the public's faith in science. The judge's power to exclude testimony is one way to minimize conflicting scientific evidence, but this solution requires criteria for determining whether testimony is truly scientific. The crisis of toxic torts and

[15] For the tie between toxic torts and concerns about expert testimony, see Tal Golan, "Revisiting the History of Scientific Expert Testimony," *Brooklyn Law Review* 73 (2008): 933. For a sympathetic discussion of asbestos manufacturers, see Peter Huber, *Galileo's Revenge* (New York: Basic Books, 1993), 148–91.

[16] Gary Edmond, "Supersizing Daubert: Science for Litigation and Its Implications for Legal Practice and Scientific Research," *Villanova Law Review* 52 (2007): 857; Huber, *Galileo's Revenge*.

[17] Alvin M. Weinberg, "Science and Trans-Science," *Minerva* 10 (1972): 209–22.

[18] Golan, "Revisiting the History of Scientific Expert Testimony."

the junk science campaign increased the drive to find adequate grounds for excluding expert testimony.

The US Supreme Court addressed this issue in *Daubert v. Merrell Dow Pharmaceuticals, Inc.* in 1993, a case regarding two children with birth defects alleged to have resulted from their mothers' use of Bendectin to treat morning sickness. Many scholars of law and science have concluded that this decision instituted a new exclusionary ethos in the judiciary, a tendency to prevent plaintiffs' scientific evidence from inclusion in court.[19] First, in *Daubert*, the Court intensified the federal judge's gate-keeping function to ensure that only valid scientific evidence would be presented. The problem with this aspect of the decision was recognized in Chief Justice Rehnquist's dissent: "I do not think [the Federal Rules of Evidence] imposes on [judges] either the obligation or the authority to become amateur scientists."[20]

Second, the court made the legal model for analyzing scientific evidence much more stringent than those used in science or regulatory hearings. In *General Electric v. Joiner*, the Court suggested that the judge should evaluate each paper, study, and piece of evidence independently rather than take a weight-of-the-evidence approach. According to this standard, the judge asks whether a particular study, on its own, proves the point at issue, rather than looking at the ensemble of evidence presented. Dissenting in *Joiner*, Justice Stevens questioned the rejection of the weight-of-the-evidence approach, which he noted is how scientific evidence is analyzed in regulatory decisions and in science itself.[21] Making each study stand alone puts a heavy burden of proof on a study. This way of analyzing studies tends to sweep most of them away, because, as discussed in Chapter 3, philosophers of science have argued for a century that it is methodologically inaccurate to think that any one study contains a crucial experiment.[22]

This consideration leads to the third problem, the adequacy of the four criteria the court suggested a judge should use to determine whether evidence represented scientific knowledge or not: (1) Can the evidence

19 Margaret Berger, "Upsetting the Balance between Adverse Interests: The Impact of the Supreme Court's Trilogy on Expert Testimony in Toxic Tort Litigation," *Law and Contemporary Problems* 64 (2001): 289–326; Edmond, "Supersizing Daubert"; Golan, "Revisiting the History of Scientific Expert Testimony"; Sheila Jasanoff, "Representation and Re-Presentation in Litigation Science," *Environmental Health Perspectives* 116 (2008): 123–9.

20 *Daubert v. Merrell Dow Pharmaceuticals, Inc.*, 509 US 579 (1993).

21 *General Electric Co. v. Joiner*, 522 US 136 (1997).

22 Pierre Duhem, *The Aim and Structure of Physical Theory* (Princeton, NJ: Princeton University Press, 1954); W. V. Quine, "Two Dogmas of Empiricism," in *From a Logical Point of View: Nine Logico-Philosophical Essays*, 2nd revised edition (Cambridge, MA: Harvard University Press, 1980), 20–47.

be tested? (2) What are the methods' error rates and standards of reliability? (3) Are the methods generally accepted? (4) Is the research peer reviewed and published? Debates over Popper's model of science suggest the problems of the first criterion, but it is the last one that is most concerning. The problem is that, because of the difficulties of trans-science and because showing that a chemical is toxic is unlikely to advance one's career, many researchers avoid these fields. There is little independent research into these public interest areas of science, so the company that produces a particular chemical generates almost all the published research on it. By rejecting research produced by plaintiffs' experts on account of its interested nature, courts have inadvertently ceded the field to manufacturer-produced research that has just as much danger of conflicts of interest. Moreover, since plaintiffs can use internal research found through discovery, this situation creates disincentives for well-meaning companies to perform research on the potential dangers of their products.[23]

Beyond internal research – and here is where the Daubert standard is central to agnotology – emphasizing publication encourages companies to prevent negative evaluations of their intellectual property from entering the scientific literature.[24] It is this fact that encourages threats to the career of the individual scientist. If a hostile article is about to appear, journal editors may be contacted, and an online campaign against it may be coordinated. Because they are largely rhetorical maneuvers to prepare the way for a call for retraction, these attacks will consist of very particular complaints about methods. Such challenges will not suggest how different methods might provide a similar result, and they might even be attacks on generally accepted methodologies. This is an odd way to do science. Generally, if you disagree with a paper's conclusions, you perform the experiments that will show that they are incorrect. Therefore, these are generally not good-faith criticisms to aid the search for scientific knowledge; they are targeted attempts to prevent papers from providing support for possible litigation or regulation.

If researchers with whom a company has a contractual relation produce threatening results, then a number of legal means help the company prevent publication. The contracts of employees and external researchers contain nondisclosure clauses that prevent the publication of negative data.[25] External researchers who use a corporate product are saddled with

[23] Edmond, "Supersizing Daubert," 905.
[24] Edmond, "Supersizing Daubert"; McGarity and Wagner, *Bending Science*.
[25] McGarity and Wagner, *Bending Science*, 109–15.

Material Transfer Agreements that can require company approval before publication.[26] While researchers are not without free will when they enter these arrangements, the contemporary context of research in many fields makes it impossible to do without such funding and materials.

If a hostile scientist is not contractually connected to a company, then he can be threatened with a libel lawsuit for criticizing a product.[27] It is not necessary to win a lawsuit to silence the scientist because merely the threat of a lawsuit silences many. Defending oneself against libel or breach of contract can be bankrupting, and universities frequently will not support the defense. Even if one perseveres in the suit, years of research will be lost, and one's reputation will be damaged.

If legal means are unavailable, then companies can use extralegal means to attack a hostile scientist to prevent her from providing future expert testimony. Given their sponsorship of academic departments and buildings, corporations can apply pressure on university officials. Corporate funders can contact a university and complain about a researcher, leading to informal discipline. Challenging the research supporting a senior colleague's biotech company can lead to poor performance reviews. Political opponents can threaten to withdraw government grants. Chairmen of congressional committees can demand data. There are many channels for informal institutional discipline.

Finally, the scientist can find her reputation attacked. Deconstructive attacks on papers can hinder a scientist's chance for journal publications and grant approvals. Her reputation can easily be attacked on the Internet or in sympathetic media outlets eager for a scandal. More troublingly, industry-backed scientists can accuse hostile scientists of scientific misconduct, leading to investigations and ethics hearings. For example, Herbert Needleman's research demonstrated the negative effects on child development of low amounts of lead.[28] Because of this research, he was attacked by the lead industry throughout his career, culminating in an ethics investigation opened by two industry-connected experts who looked at his laboratory notebooks as a result of an Environmental Protection Agency (EPA) lawsuit in which he served as a witness.[29] Although the industry-connected scientists were unable to point to any specific research

[26] Philip Mirowski, *Science-Mart: Privatizing American Science* (Cambridge, MA: Harvard University Press, 2011), 139–93.

[27] McGarity and Wagner, *Bending Science*, 168–77.

[28] Krimsky, *Science in the Private Interest*, 187–91.

[29] Herbert Needleman, "Salem Comes to the National Institutes of Health: Notes from Inside the Crucible of Scientific Integrity," *Pediatrics* 90, no. 6 (1992): 977–81.

violations, they said that they had suspicions of misconduct. These accusations led to two ethics hearings. Needleman had to rely on the University of Pittsburgh's Faculty Senate to receive an open hearing that was initially denied to him, but eventually he was completely exonerated in the second hearing. Even though he was exonerated of misconduct, he suffered reputational damage, research delays, and much anxiety.

Finally, industry actors may engage in a pattern of harassment against antagonistic scientists, as with Hayes. These attacks discredit specific researchers and their work, thus preparing the way for the exclusion of evidence by courts and regulators. Individual scientists rarely have the resources or strength to withstand such assaults. Thus, doubt is produced. To speak the scientific truth on issues central to the common good involves great personal risk. It therefore requires courage tied to the devotion to truth.

Philosophical *Parrhesia*

All of these problems should be addressed structurally by, for example, stricter conflict of interest policies or stronger defenses of scientific speech.[30] In no way do I want to downplay the importance of systematic changes for the defense of scientific integrity. Yet all such changes can have negative effects when influenced by institutional pressures. As discussed in Chapter 2, audit mechanisms are not sufficient to fix the problems of science, and the problems of agnotology show another reason why this is so. Greater attention to research integrity that focuses only on mechanisms of accountability, rather than looking to the virtues of the researcher, can give more weapons to harass scientists whose results are unpopular. Opening research to challenge can undermine the always imperfect science necessary for regulation. No simple deontological code, procedural guideline, or bureaucratic structure can completely solve these problems. The proper deployment of structural changes requires individuals with good character. Moreover, there is little political will to make such changes at the present moment, so both the defense of current science and advocacy for change requires individuals who are morally formed in such a way that they will have the courage to speak in the face of risk. Finally, as discussed in Chapter 5, any institutional system will present threats to a practice of

30 For structural solutions, see, among others, Krimsky, *Science in the Private Interest*; McGarity and Wagner, *Bending Science*; Ferric C. Fang and Arturo Casadevall, "Reforming Science: Structural Reforms," *Infection and Immunity* 80, no. 3 (March 1, 2012): 897–901.

truth. Many commentators call for an increase in government funding of basic and applied research as a way to protect against corruption by corporate money, but the prior regime of research that relied on government, mainly military, funding also presented threats to true speech, since politics has just as many interested parties as commerce. Defending scientific truth will always depend on properly formed individuals.

One could also argue that the scientific community must be reformed. Some colleagues do come to the defense of attacked scientists,[31] and many scientists speak out in public on disputed issues. However, there are too many members of the community who avoid research in politically conflicted areas or who do not assist those who are attacked out of fear for their own careers and reputations. One of the central problems for science is that the scientific ideal is the entrepreneurial subject rather than the subject of truth. Individual moral formation is at the heart of reforming the community.

One quality, or virtue, that could aid in this project is *parrhesia*. Literally, *parrhesia* means to tell everything, to speak freely, or free speech.[32] Originally, *parrhesia* was a political term that indicated a structural feature of Greek democracy.[33] Any citizen, and only a citizen, could speak freely and critically on issues before the Assembly. In philosophy, the term came to mean telling an unpleasant truth to the people, but telling such unpleasant truths came to be increasingly dangerous. In Plato's work, especially his Letters and the *Apology*, Socrates' death is tied to *parrhesia*, and the political game is seen as extremely risky for the truth-teller in an unjust city.[34]

With the downfall of Greek democracies, *parrhesia* came to be used in the context of personal spiritual guidance, with Socrates serving as the hinge of this development. The parrhesiast was the person with the skill to force one to give an account of one's life and to truthfully guide one on the road to virtue.[35] This more intimate meaning was not without political connotations, since *parrhesia* became necessary to guide the Prince.[36] If a philosopher could educate the ruler properly, this education would presumably have profound effects on the justice with which he governed his kingdom. This focus on the ruler's moral education was not just a feature of Platonic political philosophy, since individuals of many philosophical

31 E.g., Needleman, "Salem Comes to the National Institutes of Health," 981.
32 Foucault, *The Hermeneutics of the Subject*, 366.
33 Foucault, *The Government of Self and Others*, 7.
34 Foucault, *The Government of Self and Others*, 216–17; Foucault, *The Courage of Truth*, 85–90.
35 Foucault, *The Hermeneutics of the Subject*, 385.
36 Foucault, *The Government of Self and Others*, 192–6.

schools undertook the task of forming the Prince with more or, usually, less success: Aristotle with Alexander, Seneca with Nero, and Fronto with Marcus Aurelius.

The second form later political *parrhesia* took was confrontation with a ruler to intercede in the interests of mercy or justice. Peter Brown shows how the philosopher was a privileged person for interceding on behalf of his city or the powerless within it to prevent a harsh punishment or to ask for tax relief, a role that later came to reside with the bishop.[37] This political intervention nearly always included a call to conversion aimed at the ruler's way of being, so the two forms of political *parrhesia* are almost always linked.

Beyond its necessity for spiritual guidance and political intervention, *parrhesia* is central to ethics in another way. This truth is deeply held by the subject who speaks it. "The parrhesiastic enunciation is the affirmation that in fact one genuinely thinks, judges, and considers the truth one is saying to be genuinely true."[38] Rather than disinterested, objective knowledge that could be spoken by anyone, this truth is deeply affirmed by the subject who speaks it. For this to be the case, the subject must have transformed the specific *logos* into his *ethos*. It is here that *parrhesia* depends on the care of the self. The truthfulness of the speech is corroborated by the truthfulness of the life of the speaker.[39] Only the person who lives as a philosopher can speak philosophical truth with *parrhesia*.

[37] Peter Brown, *Power and Persuasion in Late Antiquity: Towards a Christian Empire* (Madison: University of Wisconsin Press, 1992).

[38] Foucault, *The Government of Self and Others*, 64.

[39] Foucault, *The Courage of Truth*, 148. The difference between this emphasis on the subjective stance toward truth and an emphasis on the propositional content of truth can be seen in MacIntyre's and Foucault's different interpretations of the Platonic dialogue *Laches*. It is a dialogue between Socrates and two Athenian generals, Nicias and Laches, about courage. To MacIntyre, the failure of the generals to define courage shows that they lacked true courage because they lacked the knowledge of what courage is (MacIntyre, *Whose Justice? Which Rationality?*, 64). Foucault argued that the dialogue is about the best way to educate children, about moral formation (*The Courage of Truth*, 122–4). Even when it deals with courage, it is about risky speech and the relation between courage and truth. Socrates is the philosophical exemplar and martyr for courageous truth-speaking, but Nicias and Laches also show courage (*The Courage of Truth*, 141–4). Unlike the interlocutors of other dialogues, such as Protagoras, Gorgias, Callicles, or Thrasymachus, who resist Socrates' style of question and answer, these interlocutors accept Socrates' method of proceeding with full knowledge that it may lead to unpleasant truths. From previous observations of Socrates, Nicias accepts that he will have to give an account of his way of life so that it might improve him (Plato, "Laches," in *Complete Works*, ed. John Cooper, trans. Rosamund Sprague (Indianapolis: Hackett, [1997], 188b). Laches accepts this risk of offense and humiliation as well, because he has seen from Socrates' actions that his discourse is authenticated by his life (Plato, "Laches," 188d–189b). The dialogue ends not with a definition of courage but with a call to continue to adopt a form of life that allows for a care of the self that fosters the courage of truth already shown by the interlocutors (Plato, "Laches," 201a–c).

Because science seeks objective knowledge, it seems an unlikely place to find *parrhesia,* with this subjective element. Yet, in figures like Galileo, Needleman, or Hayes, scientists deeply tied to their research, it does seem like these two modes of engaging truth are brought together, and these two modes of speaking the truth need to be brought together if the specific intellectual is truly to play an effective political role. Parrhesiastic discourse is necessary for the bare functioning of biomedical science under its current institutional regime. Scientists need to have the courage of truth to speak publicly about climate change, toxins, and drug trials today because they face actual threats. The expert needs to be committed to truth in her life and speak for it in the public sphere.

The quality of courageous truth-telling clearly addresses many of the problems of contemporary science indicated earlier. In the face of skewed incentives toward fabricating research or engaging in managed publishing, *parrhesia* is shown in one's life and actions as much as in one's words. Through proper care of the self and a different style of life, one binds oneself to the truth, embodying it as much as possible. Thus, the parrhesiastic scientist researches with integrity, not merely in order to follow procedural guidelines, but because it is a way of being. Because of this tie to truth, the scientist will speak it openly and fully, not allowing it to be constrained by contractual nondisclosure obligations. Of course, this truth-telling will occur only if it is timely. There is no need to think that there can be no secrets, merely no secrets that harm public safety and scientific progress. The parrhesiast speaks that truth boldly in the face of public opinion, demagogues, and doubt. The parrhesiast confronts untruth and tries to overcome it, whether it be related to tobacco and cancer, climate change, vaccines, or the latest blockbuster drug. Finally, *parrhesia* demands that one confront the risks to self, reputation, and career that such truth-telling incurs.

Given the current problems of science, it is obvious that mere training in the practices of the scientific community is not enough to develop the virtue of truth-speaking. As Chapter 5 argued, self-conscious work is necessary. The possession of *parrhesia* is the result of the care of the self and engagement in the techniques of the self. Care of the self attaches one to the truth, allowing the *logos* to be expressed in one's *ethos*, and allowing that *logos* to be spoken even in the face of risk.

The techniques discussed in Chapter 5 allow one to develop the capacity for risky truth-telling. First, meditative exercises give one the proper regard for the relative merits and value of truth and the common good versus the things risked by speaking that truth. By meditating on the brief duration of reputation in science, on the uncertain success of a career that always depends on the next grant application or the reception of the next paper, on the vanity of wealth that only gains short-term consumables, the scientist can set these goods in the proper subordinate relation to her vocation to truth and to the broader common good. A technique more directly related to this issue is the ancient *praemeditatio malorum.* One visualizes oneself with one's reputation ruined by Internet gossip and Office of Research Integrity investigations, one sees one's grants unfunded and one's laboratory closed, one imagines oneself buried in a lawsuit with no defenders. Then one exorcizes these fears by carefully showing oneself that these outcomes are not the worst that can happen, that even these will be tolerable if one adheres to virtue and truth. Through such techniques, one develops the mental constancy that would allow one to face danger.

This care of self also includes the *askesis* of one's form of life. Many of the ancient philosophical schools followed the example of Socrates and thought that the rigor of the philosophical life vouchsafed the truths of the things said by the philosopher. Christian practices of fasting, poverty, and alms-giving similarly disengage one's affections from the seeming absolute goods of this world to put them in proper perspective, while also demonstrating a commitment to the evangelical message. From such practices arises a correspondence between actions and words.

Similarly, but in a much less straightforwardly ascetic vein, the modest salary, nomadic nature, and long hours of the scientific life can prepare one to face difficulties for the sake of the truth to which one devotes oneself. Even the fictional depiction in *Arrowsmith* indicated forgoing some comforts as necessary for the scientific vocation. Steven Shapin describes how the lack of a bourgeois salary for a long time was regarded as vouchsafing the scientist's virtue and commitment to truth.[40] For the biochemist Erwin Chargaff, the poverty of a previous generation of scientists meant that only the truly motivated entered the field.[41] Adopting the style of life advocated by Sertillanges or Weber trains one through hardship to accept risk, which provides added support for the truths that the scientist advocates. The care of the self through the scientific life can be extremely valuable for helping individuals confront threats to the integrity of the practice of science.

[40] Shapin, *The Scientific Life*, 45.
[41] Chargaff, *Heraclitean Fire*, 160.

Christian Truth-Telling

Parrhesia is not merely a secular or philosophical virtue but is deeply embedded in Scripture and the practice of the early Church.[42] Yet *parrhesia* does not always have the same connotations in Christian writings as it does in Greco-Roman philosophical texts. A first aspect of bold speaking largely overlaps with non-Christian philosophical forms of speaking the truth. Thus, in the Septuagint, the personification of Wisdom boldly calls out in the streets, much as the Cynic.[43] Here, it is a matter of speaking the truths of the Wisdom literature, which largely corresponds with the Wisdom traditions of other ancient societies. Christians took up this role through preaching, and Christian bishops and ascetics assumed the social role of philosophers by interceding with the powerful for mercy and justice on behalf of the weak and their communities. According to Peter Brown, St. Ambrose's intervention with Emperor Theodosius after a massacre at Thessalonica takes the form not of pastoral power but of bold philosophical speech.[44] The Christian bishop carried out the role of an ancient spiritual guide pointing to the dangers of anger and its cure in penance.

Christianity adds to this wider definition by seeing *parrhesia* in the proclamation of the gospel. *Parrhesia* and its forms appear throughout the Acts of the Apostles in relation to apostolic preaching, and Paul uses such terms when describing his activity and when others describe his ministry.[45] In the books of Maccabees and in later Christian writings, it is the martyrs who primarily display the courage of truth both in their defense of the faith during interrogations and by their death whereby they witness to the truth of the faith with their blood.[46]

[42] This discussion of *parrhesia* in Christianity draws on Foucault, *The Courage of Truth*, 326–38; Heinrich Schlier, "Παρρησία, Παρρησιάζομαι," ed. Gerhard Kittel and Gerhard Friedrich, *Theological Dictionary of the New Testament* (Ann Arbor, MI: Eerdmans, 1967); G. J. M. Bartelink, "Quelques observations sur parrēsia dans la littérature paléo-chrétienne," in *Graecitas et Latinitas Christianorum Primaeva. Supplementa; Fasciculus 3* (Nijmegen: Dekker & Van de Vegt, 1970), 7–57; Brown, *Power and Persuasion in Late Antiquity*; W. C. van Unnik, "The Christian's Freedom of Speech in the New Testament," *Bulletin of the John Rylands Library* 44 (1962): 466–88; Stanley Marrow, "Parrhesia and the New Testament," *The Catholic Biblical Quarterly* 44 (1982): 431–46. Craig Hovey addresses *parrhesia* from the perspective of contemporary Christian ethics, but he primarily embeds Christian witness in a juridical framework, rather than one of politics and care of the self. He also refuses to engage ancient nontheological sources. See Craig Hovey, *Bearing True Witness: Truthfulness in Christian Practice* (Grand Rapids, MI: Eerdmans, 2011).

[43] Prv 1:20–1.

[44] Brown, *Power and Persuasion in Late Antiquity*, 111–12.

[45] Schlier, "παρρησία, παρρησιάζομαι," 882–3. For a discussion of *parrhesia* in Paul in relation to Jewish sources, see Michael Cover, *Lifting the Veil*, (Berlin: De Gruyter, 2015), 271–80.

[46] Schlier, "παρρησία, παρρησιάζομαι," 885.

This truth-speaking of the apostles and martyrs images God's speaking of truth about Himself. In the Septuagint, God shows Himself in Psalm 94:1–3, where *parrhesia* translates a Hebrew term that means "shines forth" and is used for God's appearances in other places.[47] In John's gospel, the bold preaching of the good news begins with Jesus himself. While full revelation of his identity and gospel will await the sending of the Paraclete,[48] Jesus speaks of himself in open public proclamation: "I have spoken openly to the world; I have always taught in synagogues and in the temple, where all the Jews come together. I have said nothing in secret."[49] Thus, human truth-speaking is the reflection of divine self-revelation and the communication of the *Logos*.

The courageous speaking of the truth to other people is supported by the Christian's friendship with God. In Hellenistic thought, truly virtuous friends are able to speak boldly and openly to one another.[50] Similarly, through the divine filiation in which Christians become children of God in baptism, Christians become friends of God and can thus speak to God confidently.[51] Christian *parrhesia* is a distinct mode of being that is characterized by a trust and confidence in God that is primarily experienced in prayer. Foucault calls this mode vertical *parrhesia* toward God in contrast to the horizontal *parrhesia* toward other people.[52]

This understanding of the tie between speaking the truth and friendship with God was already developed in the Septuagint, Philo, and Josephus. In these sources, to speak openly to God is a privilege of the righteous,[53] so there is a tie between moral character and this boldness of friendship. This boldness is expressed through one's confident prayer to the Lord.[54] In the New Testament, especially in Hebrews, this confidence and trust in God flows from the Christian's relationship to Jesus.[55] It is because of Christ's

[47] Dt 33:2, Ps 49:1–3.

[48] Jn 16:23.

[49] Jn 18:20. Scriptural citations from Harold Attridge, ed., *HarperCollins Study Bible: New Revised Standard Version* (San Francisco, CA: HarperOne, 1989).

[50] For the relation between friendship and frank speech in Greek and Christian writings, see John T. Fitzgerald, ed., *Friendship, Flattery, and Frankness of Speech: Studies on Friendship in the New Testament World* (New York: Brill, 1996).

[51] Bartelink, "Quelques observations sur parrēsia dans la littérature paléo-chrétienne," 13. These ties between *parrhesia*, filiation, and baptism may be visible in the Antiochean baptismal rite. See R. G. Coquin, "La theme de la parrhesia et ses expressions symboliques dans les rites d'initiation a Antioche," *Proche Orient Chretien* 20 (1970): 3–19.

[52] Foucault, *The Courage of Truth*, 326–7.

[53] Schlier, "παρρησία, παρρησιάζομαι," 876–8; Bartelink, "Quelques observations sur parrēsia dans la littérature paléo-chrétienne," 10–11.

[54] 1 Jn 5:14.

[55] Hb 3:6.

sacrifice on the Cross, in which Christ becomes an intercessor for the Christian, that the Christian can approach the throne of God with confidence.[56] This trust in God allows a freedom and boldness in prayer. The righteous can complain and plead to God, and the saints can intercede with God for the Church on earth through their ties of friendship.[57]

This confidence in God undergirds the Christian's boldness toward other people in proclaiming truth and the gospel:

> *Parrhesia* is not just the courage one demonstrates in the face of persecution in order to convince others, [but also a] courage [which] is confidence in God, and this confidence cannot be separated from one's courageous stance towards others. What distinguishes the courage of someone like Socrates, or Diogenes, for example, from the martyr's courage . . . is precisely that the former is only the courage of man addressing other men, whereas the courage of the Christian martyrs rests on this other aspect, this other dimension of the same *parrhesia*, which is trust in God; confidence in salvation, in God's goodness, and also in his listening.[58]

Trust in God's Providence allows the Christian to serve as an instrument of that Providence, by witnessing to one's faith and by speaking on behalf of justice and the common good.

How does the Christian gain *parrhesia*? Linked as it is to charitable friendship with God and the proclamation of faith, it can only be a gift of God through grace. It is given by the Holy Spirit.[59] The early Christians prayed,

> "And now, Lord, look at their threats, and grant to your servants to *speak your word with all boldness* . . ." When they had prayed, the place in which they were gathered together was shaken; and they were all filled with the Holy Spirit and *spoke the word of God with boldness.*[60]

Parrhesia is thus a gift to be asked for through prayer. John Chrysostom and Theodore of Mopsuestia also claim that *parrhesia* is given in baptism.[61] The Christian is adopted by God through God's free action by which he is taken up into the body of Christ, and thus receives the ability to boldly speak truth to God and man.

Yet this gift from God requires a free human response in the Christian life. The perseverance of martyrdom in Old Testament and Christian texts

[56] Hb 4:14f, 10:19.
[57] Bartelink, "Quelques observations sur parrēsia dans la littérature paléo-chrétienne," 25.
[58] Foucault, *The Courage of Truth*, 332.
[59] Marrow, "Parrhesia and the New Testament," 443–4.
[60] Acts 4:29–31. Emphasis added.
[61] Coquin, "La theme de la parrhesia et ses expressions symboliques dans les rites d'initiation a Antioche," 5.

both expressed *parrhesia* and gained the martyr a special friendship with God.[62] Already in the Old Testament and Philo, it was the righteous and those who lived according to wisdom who possessed *parrhesia*.[63] This tie to ethics continues in the New Testament:

> Let us love, not in word or speech, but in truth and action. And by this we will know that we are from the truth and will reassure our hearts before him whenever our hearts condemn us; for God is greater than our hearts, and he knows everything. Beloved, if our hearts do not condemn us, we have *boldness* before God; and we receive from him whatever we ask, because we obey his commandments and do what pleases him.[64]

Patristic texts, in different ways, showed *parrhesia* resulting in part from self-work. In the monastic tradition, the tests of temptations and *askesis* were viewed as a kind of everyday martyrdom through which perseverance in virtue deepened one's *parrhesia*.[65] Basil shows the importance of asceticism for *parrhesia* in his confrontation with an anti-Nicene praetorian prefect, where he argues that confiscation, exile, torture, and death are nothing to him because he owns little, considers himself a guest of God in the world, and would see death as a gain in seeing God sooner.[66] This attitude allows him to freely defend the faith in the face of imperial power. In the mystical tradition, contemplative exercises that assisted the mystical ascent to union with God were viewed as deepening the Christian's friendship with and confidence in God and thus leading to greater *parrhesia*.[67] Though *parrhesia* is God's gift, Christian action, devotion, and spiritual exercises can deepen and strengthen it. Speaking the truth requires living and embodying the truth, which can be accomplished only through the care of the self with the help of grace.

Risky Speech for the Christian in Science

What does Christian truth-telling add to the scientific devotion to truth discussed earlier? In one sense, bold Christian speech will reinforce the fight against the degradation of scientific practice through commercial or personal interests because it shares many aims with it. The Christian

62 Bartelink, "Quelques observations sur parrēsia dans la littérature paléo-chrétienne," 25.

63 Schlier, "παρρησία, παρρησιάζομαι," 876–8; Bartelink, "Quelques observations sur parrēsia dans la littérature paléo-chrétienne," 10–11.

64 1 Jn 3:18–22. Emphasis added.

65 Bartelink, "Quelques observations sur parrēsia dans la littérature paléo-chrétienne," 26.

66 Gregory Nazianen, "On St. Basil the Great, Bishop of Caesarea," in *Funeral Orations*, trans. Leo McCauley, The Fathers of the Church, v. 22 (New York: Fathers of the Church, 1953), 49.

67 Bartelink, "Quelques observations sur parrēsia dans la littérature paléo-chrétienne," 28–9.

embraces all searches for truth, even in the natural realm, since Jesus is the *Logos* through whom the natural order was created. Realizing that Creation reveals in some way its Creator gives the integrity of scientific practice an added importance for Christians. Christianity also adds resources for bolstering courage. The Crucifixion and the history of martyrs reveal the risks of speaking any truth in a fallen world. Meditating on and using such exemplars as models bolsters courage. The Christian can also have ultimate confidence in God's victory and thus has no need for fear. Moreover, Christian devotional practices involve the techniques of the self invoked by secular philosophers. In these ways, Christianity can strengthen one's will to protect the practice of science.

In another way, the broader truth in which Christian practices form one can oppose the dangerous tendencies of a reductionist rationality through broader normative commitments. Such a curtailed reductionist worldview threatens to encourage the treatment of others as material for use and thus to subordinate the weak to powerful. In contrast, the Christian remains committed to the common good. The embodying of Christian truth through the imitation of Christ counteracts negative aspects of scientific formation because it allows one to see the Providence active in nature and to see the image of God in other people. Thus, people and objects cannot merely be dissolved into a network of relations to be manipulated.

In the last century, the most obvious manifestation of this subordination of humans to social need was eugenics under its various guises, from state-enforced negative eugenics to the subtler eugenics of the entrepreneurial family. Catholics were at the forefront of a broad social opposition to the earlier coercive regime of sterilization.[68] Yet the most important resistance to this negative eugenics program was that of scientists such as J. B. S. Haldane, Julian Huxley, Lancelot Hogben, and Herbert Jennings.[69] They fought eugenics by changing the scientific paradigm to make negative eugenics seem nonscientific. Their opposition was driven not only by scientific considerations but also by their leftist political views, which drove them "to recognize that mainline eugenics expressed race and class prejudice."[70] Similarly, other Marxists, Stephen J. Gould and Richard Lewontin, opposed the genetic determinism of sociobiology on scientific grounds.[71] In a memorial to Gould, Lewontin and Richard Levins compare Gould's life as a public intellectual to Haldane's, and saw his

[68] Sharon Leon, *An Image of God: The Catholic Struggle with Eugenics* (Chicago: University of Chicago Press, 2013); Kevles, *In the Name of Eugenics*, 118–19.
[69] Kevles, *In the Name of Eugenics*, 122–8.
[70] Kevles, *In the Name of Eugenics*, 127.
[71] Kevles, *In the Name of Eugenics*, 84–5.

public advocacy and popular scientific work as driven by radicalism.[72] Beyond moral considerations, the specific intellectual can oppose injustice through science itself.

It is unfortunate that it is largely atheist socialists rather than Christians who have undertaken the fight against dangerous deployments of power in genetics through the means of science. But Marxists, like Christians, are formed in a certain way of life with a devotion to a truth and an eschatological horizon, even if only an immanent one. These aspects form them in a manner beyond the mere reductionist rationality of science, allowing them to see when science is allowing the normative presuppositions of contemporary power structures to direct its research.[73] These examples show that Christians could also deploy risky speech in a distinctive way in science.

The Christian who is a scientist has another need for bold speech. Because of portrayals of the incompatibility of science and faith since the Enlightenment, religious scientists have frequently faced hostility for their beliefs. For this reason, courage is necessary for the scientist to proclaim the compatibility of faith and reason. For example, Pierre Duhem, a physicist, philosopher, and historian who died in 1916, was never called to take an academic position in Paris partly because of hostility to his Catholic faith in the secular Third Republic, so he spent his life in provincial academic centers.[74] He also saw himself forced to defend his philosophy of science against charges that it was based in faith.[75] More recently, Francis Collins, a respected geneticist who managed the Human Genome Project, faced opposition to his appointment as director of the National Institutes of Health because he had written a best-selling book on the compatibility of science and Christianity and started a foundation that researched the relations between science and religion.[76]

[72] Richard Lewontin and Richard Levins, "Stephen Jay Gould – What Does It Mean to Be a Radical?," in *Stephen Jay Gould: Reflections on His View of Life*, ed. Patricia Kelley and Robert Ross (New York: Oxford University Press, 2009), 199.

[73] Of course, Marxism might not be a critical resource in a Communist society.

[74] R. N. D. Martin, *Pierre Duhem: Philosophy and History in the Work of a Believing Physicist* (La Salle, IL: Open Court, 1991).

[75] Robert J. Deltete, "Man of Science, Man of Faith: Pierre Duhem's 'Physique de Croyant,'" *Zygon* 43, no. 3 (2008): 627–37.

[76] Francis S. Collins, *The Language of God: A Scientist Presents Evidence for Belief* (New York: Free Press, 2007); Jerry Coyne, "Francis Collins as NIH Director," *Why Evolution Is True*, https://whyevolutionistrue.wordpress.com/2009/07/10/francis-collins-as-nih-director/; "Steven Pinker on Francis Collins," *Why Evolution Is True*, http://whyevolutionistrue.wordpress.com/2009/07/11/steven-pinker-on-francis-collins/; Chris Wilson, "Jesus Goes to Bethesda," *Slate*, July 9, 2009, www.slate.com/articles/health_and_science/science/2009/07/jesus_goes_to_bethesda.html.

Proclaiming the rationality of the gospel can have serious repercussions, but this task is sorely needed.

In the other direction, scientists need courage to confront their faith communities with the need to accept scientific findings. The Catholic Magisterium has shown reluctance in the past to engage in the parrhesiastic game with those advocating a reconciliation with contemporary forms of scientific research. The classic example, of course, is Galileo.[77] Galileo advocated what he saw as the true structure of the world against those like Cardinal Bellarmine who followed their interpretation of Aquinas and Aristotle by taking an instrumentalist approach toward astronomical knowledge, an advocacy for which Galileo suffered.[78]

The Catholic Church has admitted that its handling of the Galileo affair was mistaken, and later approaches to controversial scientific advances have been more cautious. For example, the theory of evolution was never condemned, but specific works and authors such as John Zahm and Dalmace Leroy arguing for different ways of integrating evolution and doctrine were investigated by the Congregation of the Index.[79] In the case of Zahm, the Congregation was suspicious of his argument that the human body, but not the soul, could be the result of evolution.[80] Yet the Church did not want to take a firm position condemning evolution, so, with the help of his superiors, Zahm worked out a resolution in which his book was never officially condemned, even though publication of it was withdrawn.[81] In the end, his work would approximate the Church's later way of integrating Genesis and evolutionary theory. Here the parrhesiastic game was not totally successful, but over time, the Catholic Magisterium did accept these conclusions, even though the scientist suffered for his stand.

Yet caution can be warranted on the part of the Church. Science is not always clear, and it can be wrong. Its claims must be balanced against the understandings of Christian anthropology when they are related to the basic structure of the human person. As Foucault teaches us, the

[77] Although one could also accuse Galileo himself of lacking the skill necessary for successful *parrhesia* in that he needlessly alienated his benefactor, Pope Urban VIII, through his mode of writing.

[78] For alternative interpretations of instrumentalism and truth in this debate, see Duhem, *To Save the Phenomena*; MacIntyre, "Epistemological Crises and Dramatic Narrative."

[79] Mariano Artigas, Thomas F. Glick, and Rafael Martínez, *Negotiating Darwin: The Vatican Confronts Evolution, 1877–1902* (Baltimore, MD: Johns Hopkins University Press, 2006).

[80] John Zahm, *Evolution and Dogma* (Chicago: D. H. McBride, 1896), 340–68.

[81] Zahm's case, like many others, was involved in other ecclesial debates, which were often more important for the investigation than the issue of evolution. In his case, his book was read through the lens of the turn-of-the-century conflict over Americanism.

knowledge of the human sciences is always related to contemporary power structures, if not reducible to them, so the Church must ensure that it does not equate transitory contemporary social life with the essence of the human person. Thus, the Church has opposed genetic determinism, negative eugenics, dominance hierarchies, reduction of human action to social structures, and neural reductionism. In these cases, it seems to be on the right side of ethics and the history of science. Care is thus warranted, but it does not relieve Church authorities of the duty to listen to the unfamiliar truths for which faithful scientists argue or for the Christian theologian and scientist to engage in risky truth-telling toward the Church.

Conclusion

The current structure of science makes the researcher vulnerable to many external pressures. These threats require the scientist to undertake a political role at great personal and professional risk in order to defend her practice and society. This calling would be greatly aided if she gains the virtue of *parrhesia* by crafting herself into a subject of truth through techniques of the self. Christian truth and practice, by giving a larger context of truth and devotion to the common good, can support this political form of scientific truth-telling. It also gives the scientist other duties. She must evaluate scientific theories and policies by moral commitments that frequently challenge the normative commitments of the contemporary form of biopower to which the human sciences are linked. Faith can force the scientist to confront her scientific community with her religious beliefs and her religious community with her scientific knowledge. Both confrontations are risky. However, faith provides more resources than secular care of the self to face these risks, since bold speech is the gift of the Holy Spirit expressing the relationship of divine filiation. Truth-speaking and the truth spoken are both gifts.

This analysis raises a number of questions about the ties between truth, subjectivity, and Christian anthropology, questions alluded to throughout this book, but inadequately addressed up to this point. What is this truth that transforms the subject? In one reading of postmodern thought, it could be mere relativism and nihilism, which, if accurate, would destroy this model of ethics. What is this subject who is transformed? The malleability of subjectivity found in some forms of social theory threatens the possibility of finding any stable mode of existence. In Chapter 7, I will explore these questions by comparing the models of the subject and truth discussed so far with Augustinian conceptions of the *imago dei*.

CHAPTER 7

Subjectivity, Truth, and Theological Anthropology

Despite their connection to truth-telling discussed in the last chapter, many commentators remain skeptical of these practices of the self. First, they see the turn to practice as a rejection of truth. This opinion is reinforced when some advocates for the care of the self celebrate the idea that these practices have no need for dogmatic truth claims. Further, describing science in terms of sets of socially determined practices that lead to subjectively embraced truth claims seems to make the objective truth of science impossible. There is a danger that natural science will be reduced to a culturally relative practice or, worse, a mere exercise of power. These claims were at the heart of the science wars of the 1990s.[1] Second, critics suspect that if one makes the subject too malleable to such practices, then the possibility of any kind of solid grounding to human identity will disappear. Practice theory seems to threaten to dissolve the individual into the play of masks, discourses, or, in a more scientistic vein, memes. It is very difficult to imagine Christian ethics existing with neither truth nor continuity of the subject, which suggests serious problems for using practice theory as a tool for exploring the relationship between religion and science.

The first part of this chapter addresses these concerns, showing that accepting the analyses discussed in this book requires neither relativism nor a denial of the existence of a subject. Instead, a focus on practices of the self merely directs our attention away from the objective content of truth claims toward the ways in which the person prepares herself to pursue truth. It also focuses on the transformations that truth can work on

[1] There are many works that contributed to or analyzed the science wars, but for a good collection discussing the Sokal hoax that was at the center of these debates, see Editors of Lingua Franca, ed., *The Sokal Hoax: The Sham That Shook the Academy* (Lincoln: University of Nebraska Press, 2000).

subjectivity. This change of focus is not a denial of the objectivity of those truth claims; it is merely a new avenue of investigation. Further, rather than positing a universal model of the subject, this focus on practices related to truth emphasizes the role of conversion between different forms of subjectivity. This analysis of the transformative potential of truth is especially valuable for Christian ethics because it gives theologians tools to more deeply engage with the ethical changes wrought by faith.

Critics are correct that some variants of practice theory have certain failings from a Christian perspective. Though social theoretical work is extremely valuable for providing a formal model of ethics, it does not provide a substantive ethic. Even in the terms of Foucault's own understanding of ethics, a substantive ethic would require an ethical *telos*, an actual model of the subject to serve as a moral exemplar, and a mode of veridiction, a system for judging truth claims. It is unclear that a social theory of practices can deliver either on its own. In contrast, Christianity provides both a system of the truth of the world and a model of the subject in Christ, who is both *Logos* and image of God. Yet, while accepting the need for an underlying rationality of the world, theologians must also recognize the validity of the insights of social theory on the historicity and fallibility of contemporary claims to truth. MacIntyre's rationality of traditions helps to recover this promise of a true vision of the world while recognizing these insights.

Even without a concrete form of life as an ethical *telos*, the focus on truth and subjectivity found in practice theory is still extremely helpful for analyzing Christian theological anthropology. Perhaps the theological vision of the person that makes one most aware of the ties between truth and the subject is the Augustinian analysis of the *imago dei*. This analysis traces the intricate links between knowledge of self, knowledge of God, and practical action, arguing that one must grow in knowledge of God to fully actualize one's nature. The last section of this chapter addresses how the formal understanding of the subject described in this book fits with this concrete anthropology. On the one hand, this substantive vision corrects and perfects some aspects of social theory that may be troubling for moral theologians. On the other hand, the use of practice theory helps to address criticisms of the Augustinian focus on the self and individual desire by showing the social nature of the self and by shifting the target of ethical work from desire to practice. In this way, secular ethics serves as a helpful tool for theological analysis. Before turning to this Christian response to the problem of the subject, though, it is necessary to explore issues of truth and subjectivity more deeply.

Truth and Practice

Chapters 3 and 4 discussed two models of the connection between character and conceptual system, both of which rejected the fact–value distinction. For MacIntyre, traditions of thought are tied to certain kinds of social structures, and these traditions and structures shape both the understanding of and the formation in virtue. For Foucault, power and knowledge are intimately interrelated, with power providing sites for the development of knowledge and disciplines of knowledge reinforcing power by constructing new objects for it to grasp. These interlocking systems of power and knowledge are the sources of what Foucault calls subjectivation, the process of social formation that gives shape to one's subjectivity.

These social foundations of the grasp of truth might seem to undermine the importance of truth for an ethical subjectivity. Many philosophers argue that this seeming relativism undermines the normative import of Foucault's project.[2] Moreover, such a position toward truth seems especially troubling for the ethics of science or philosophy, which embraces devotion to truth as a foundational virtue. If truth is only an effect of power, then this ethics makes no sense. Some of these concerns can be dissipated by placing practice theory in the French tradition of the philosophy of science, a tradition exemplified by Gaston Bachelard, Jean Cavailles, and Georges Canguilhem.[3] While this tradition respected science as one of the highest human pursuits and embraced its increasing formalization, these thinkers also recognized that scientific rationality is one of the central contemporary intellectual problems.[4] Like Kuhn and

[2] For example, Charles Taylor argues that Foucault ultimately needs substantive conceptions of freedom and truth; Charles Taylor, "Foucault on Freedom and Truth," in *Foucault: A Critical Reader*, ed. David Hoy (New York: Basil Blackwell, 1986), 69–102. Maria Antonnacio rejects Foucault and points to Taylor's ethics as a solution; Maria Antonaccio, "Contemporary Forms of Askesis and the Return of Spiritual Exercises," *Annual of the Society of Christian Ethics* 18 (1998): 69–92. Similar criticisms of the lack of clear normative guidelines from the perspective of feminist and political thought come from Nancy Fraser and Michael Walzer: Nancy Fraser, "Foucault on Modern Power: Empirical Insights and Normative Confusions," *Praxis International*, no. 3 (1981): 272–87; Michael Walzer, "The Politics of Michel Foucault," in *Foucault: A Critical Reader*, ed. David Hoy (New York: Basil Blackwell, 1986), 51–68.

[3] Michel Foucault, "Introduction," in *The Normal and the Pathological*, by Georges Canguilhem (New York: Zone Books, 1989), 7–24; Michel Foucault, *The Government of Self and Others* (New York: Palgrave Macmillan, 2010), chap. 1. For further background on French philosophy of science, see Gary Gutting, *Michel Foucault's Archaeology of Scientific Reason: Science and the History of Reason* (New York: Cambridge University Press, 1989); Walter Privitera, *Problems of Style: Michel Foucault's Epistemology* (Albany: State University of New York Press, 1995).

[4] Foucault, "Introduction," 11.

Polanyi, these thinkers saw that the history of science cannot be characterized as a continual progressive development but is instead constituted by ruptures and revolutions.

The problem of the relation between social structure and knowledge becomes especially pressing when one turns from theoretical disciplines of natural science like physics or mathematics to the human sciences, like psychology, economics, or medicine. These disciplines raise problems distinct from other forms of science because, in them, knowledge and power are inherently interrelated due to the fact that these human sciences are tied to broader social practices, technologies, and institutions from which they cannot easily be dissociated.[5] Changes in medical or psychological guidelines have an immediate impact on patients' lives. These disciplines cannot but be tied to social practices and so cannot gain the level of autonomous formalization found in other scientific disciplines, but this does not mean that the objects studied by these sciences do not exist.[6] Neither do these studies of the social construction of a discipline deny that discipline's validity:

> One can show ... that the medicalization of madness ... was connected with a whole series of social and economic processes at a given time, but also with institutions and practices of power. This fact in no way impugns the scientific validity of the therapeutic effectiveness of psychiatry: it does not endorse psychiatry, but neither does it invalidate it.[7]

While, in a way, these disciplines and their objects are fabricated by contemporary social practices and practices of knowing them, fabricated in the sense of shaped or generated, they still exist. It is what C. G. Prado classifies as a "tacit realist" position, one that accepts the existence of social institutions and the objects of scientific study but does not develop an account of what such existence means.[8]

Yet even if this focus on practices of seeking truth does not deny the knowledge discovered by these practices, there are some important distinctions between this model and more conventional notions of contemporary

[5] Note that Foucault never questions the truth or validity of mathematics, physics, or chemistry. To him, these disciplines have reached a stage of scientificity in which they escape from the most important interrelations with power and develop on their own trajectories of increasing formalization. See Foucault, *Discipline and Punish*, 227.

[6] Ian Hacking, *The Social Construction of What?* (Cambridge, MA: Harvard University Press, 2000); Michel Foucault, "The Ethics of the Concern for Self as a Practice of Freedom," in *Ethics, Subjectivity, and Truth*, ed. Paul Rabinow (New York: New Press, 1998), 297.

[7] Foucault, "The Ethics of the Concern for Self as a Practice of Freedom," 296.

[8] C. G. Prado, *Searle and Foucault on Truth* (New York: Cambridge University Press, 2006), 96–8.

scientific approaches to truth deriving from the Cartesian method. The latter forms of truth-seeking aim at objectivity, attained through standard procedures that anyone can perform. In this conventional model, the use of proper method allows one to attain truth in any field.

There is another model of truth-seeking in the Western tradition. This model depends on the preparation of the subject who seeks truth, on a specific event of truth, and on a certain relation to truth. It is a truth "linked to the opportunity, to the *kairos*, and must be seized ... The agents of this discontinuous truth are those who possess the secrets of times and places, those who undergo tests of qualification, ... and those again whom truth has chosen to sweep down on." It is "a truth that occurs as an event," which echoes ideas found also in de Certeau and Badiou.[9] As discussed in Chapter 5, ancient and medieval ideas of philosophy similarly saw the acquisition of knowledge as a process by which one exercises oneself in order to become the type of person who can achieve truth. This truth then further transforms one's life. Since such truth can be grasped only at the right moment or in the right situation, attaining truth is not the work of the individual alone, but also requires an element of receptivity.[10]

This second model of truth may seem irrationalist or subjectivist, and thus distant from any conception of science. However, I would argue that at least some aspects of the scientific life could be interpreted in light of such an analysis. For example, the young scientist must undergo training and apprenticeship to be capable of engaging scientific truth. This idea is not much different from what is found in MacIntyre, Kuhn, or Polanyi, although their analysis is more clearly developed.

An even more apt example is the moment of scientific insight. A scientific discovery gives a thrill and exultation that comes from gaining an insight into the workings of nature, into the character of reality. As biologist Robert Pollack notes, every scientist works in order to have these moments of insight.[11] Indeed, Weber argues that without such moments, a person has no chance of being a good scholar.[12] Weber describes many

[9] Michel Foucault, *Psychiatric Power: Lectures at the College de France 1973–1974*, ed. Arnold I. Davidson and Graham Burchell (Basingstoke: Palgrave Macmillan, 2006), 237; Michel de Certeau, *The Practice of Everyday Life* (Berkeley: University of California Press, 2011); Alain Badiou, *Saint Paul: The Foundation of Universalism* (Stanford, CA: Stanford University Press, 2003).

[10] Michel Foucault, *Security, Territory, Population: Lectures at the Collège de France, 1977–78*, ed. Michel Senellart, trans. Graham Burchell (New York: Palgrave Macmillan, 2007), 212–13.

[11] Robert Pollack, *The Faith of Biology and the Biology of Faith* (New York: Columbia University Press, 2000), 13.

[12] Max Weber, "Science as a Vocation," in *From Max Weber*, ed. H. H. Gerth and C. Wright Mills (New York: Oxford University Press, 1958), 136.

techniques that people use to try to encourage insight to come, such as smoking a cigar on a sofa or walking up a street, but none is sure to succeed because the scientist is not in control of these moments of insight. Of course, without the hard work of generating and analyzing data, the ground will not be prepared for these insights. Efforts are necessary prerequisites for grasping a new truth, but they are not sufficient.

For this reason, scientific insights are often compared to religious experiences.[13] Polanyi has gone further than most in analyzing scientific discovery as mystical experience. For him, contemplation of nature and specifically of one's data is a way of breaking free from oneself, of transcending oneself, by breaking free of one's conceptual framework. "Scientific discovery, which leads from one such framework to its successor, bursts the bounds of disciplined thought in an intense if transient moment of heuristic vision ... The most radical manifestation of this urge to break through all fixed conceptual framework is the act of ecstatic vision."[14]

Because it originates in such semi-religious experiences, for Polanyi this knowledge becomes personal knowledge, the object of a kind of faith.[15] The scientist becomes devoted to the new idea because, for that scientist, this idea represents reality: "We accept [a theory] in the hope of making contact with reality."[16] Because of the experience of insight, the scientist ties himself to the truth, which connects Polanyi's analysis to the notion of *parrhesia* because it is this personal devotion to truth that drives one to speak truthfully in risky situations. Since a scientist engages in an approach to a fuller truth that will never be completely comprehended, a truth that is only a regulative ideal accepted in faith, scientific knowledge has many similarities to religious knowledge, especially the eschatological character of Christian truth. There is a constant inbreaking of insight about the shape of reality, but also an understanding that a final grasp of this reality is a distant hope. In both, this experience of reality and faith in the coming of a fuller understanding are keys to personal transformation.

MacIntyre's discussion of Ludwig Boltzmann's suicide shows this personal aspect of knowledge:

> The history of science is a complex set of dramatic narratives in which individuals are at key points forced to gamble on future outcomes ... When Boltzmann committed suicide, partly ... because of the growing skepticism ... about the kinetic theory, he behaved appropriately as a man who

[13] Pollack, *The Faith of Biology and the Biology of Faith*, 16.

[14] Michael Polanyi, *Personal Knowledge: Towards a Post-Critical Philosophy* (Chicago: University of Chicago Press, 1962), 196.

[15] Polanyi, *Personal Knowledge*, 266.

[16] Polanyi, *Personal Knowledge*, 5.

> had staked his all on the realism of the kinetic theory, on the possibility of carrying through that moral task which he had made central to his own life.[17]

Yet MacIntyre rejects Polanyi's description of insight as a conversion experience because he fears that it would tend toward irrationalism.[18] MacIntyre, like Kuhn, never investigated the originating instances of a scientific theory or a tradition, so traditions become merely givens to be negotiated by great minds.

MacIntyre's most extensive discussion of conversion occurs when, in discussing Edith Stein's conversion to Catholicism, he mentions two other conversions by Jewish intellectuals, Franz Rosenzweig's reengagement with Judaism and Gyorgy Lukács's conversion to Marxism.[19] His emphasis in the first two cases is on continuity in both belief and character. Rosenzweig's "conversion was not a moment of discontinuity in his life, but rather the discovery of a continuity" through his deeper engagement with a concrete community.[20] His narrative of the conversion of Stein, who was an atheist at the time of her conversion, provides almost a perfect Thomistic case of grace perfecting nature. "She was given no reason by her conversion to move away from any of the fundamental philosophical positions that she had adopted."[21] Instead, conversion provided new ways forward on questions that her philosophical convictions had raised. It is only Lukács's conversion to proletarian practices that is transformative, "a recognizable Kierkegaardian 'Either-Or,' a choice so fundamental that it cannot be supported by reasons, since it is a choice that determines what is henceforward going to count as a reason for the one who makes the choice." Lukács would later argue, however, that Marxism was "the rational solution of his earlier philosophical difficulties," and disown

[17] Alasdair MacIntyre, "Objectivity in Morality and Objectivity in Science," in *Morals, Science, and Sociality*, ed. H. Tristram Englehardt and Daniel Callahan (Hastings-on-Hudson, NY: Hastings Center, 1978), 37.

[18] Alasdair MacIntyre, "Epistemological Crises and Dramatic Narrative," in *The Tasks of Philosophy: Selected Essays*, vol. 1 (Cambridge: Cambridge University Press, 2006), 17.

[19] Alasdair MacIntyre, *Edith Stein: A Philosophical Prologue, 1913–1922* (Lanham, MD: Rowman & Littlefield, 2006), 143–61. He also discusses Adolf Reinach's conversion to Christianity in this chapter, but in much less detail. There is a general problem with MacIntyre's discussions of conversion, as they tend not to focus on grace. There is only a brief discussion of conversion through grace in MacIntyre's writings. See Alasdair MacIntyre, *Whose Justice? Which Rationality?* (Notre Dame, IN: University of Notre Dame Press, 1988), 157, 181. See also the discussion of the problems with MacIntyre's account of conversion in David Fergusson, *Community, Liberalism and Christian Ethics* (New York: Cambridge University Press, 1999), 133.

[20] MacIntyre, *Edith Stein*, 169.

[21] MacIntyre, *Edith Stein*, 174.

"Kierkegaardian irrationalism."[22] A Kierkegaardian understanding of conversion seems to generate a dangerous instability in the subject, as MacIntyre's earlier writings on Lukács describe "a recurrent instability in which the would-be settled views of any one period in his life are undermined by his insights at other periods."[23] He criticizes Lukács's unstable character: "the face behind the mask has taken on the aspect of a mask."[24] Because of these concerns about irrationalism, MacIntyre underrates the transformative possibilities of thought, leaving social practices as the only agent in the story. Foucault and Polanyi, because of their focus on this experience of insight, see how new conceptions of reality drive a process of conversion in the subject and lead to a change in the world.

One can prepare for such events of truth by critically raising problems about an issue through genealogy. Understanding the role of history and social practices in knowledge and behavior allows us to "separate out, from the contingency that has made us what we are, the possibility of no longer being, doing, or thinking what we are, do, or think."[25] By learning the historical and social production of the entrepreneurial subject, we can know that we are not doomed to these categories and that we can change our social practices. There is always an outside to systems of thought, in the voices of the marginalized, different traditions, experimental evidence, or the work of thought. It is thought from the outside that allows for change, and it is the event of thought that allows for transformation.[26]

Yet, as Polanyi and MacIntyre emphasize, these insights are always mediated through a community. Even the very process of a discovery is shaped by the milieu of the laboratory. Discussions with colleagues form inchoate ideas into clear concepts. As Ludwik Fleck notes in regard to scientific theories,

> thoughts pass from one individual to another, each time a little transformed, for each individual can attach to them somewhat different associations ... After a series of such encounters, practically nothing is left of the original content ... After making several rounds within the community, a finding often returns considerably changed to its

[22] MacIntyre, *Edith Stein*, 160.

[23] Alasdair MacIntyre, "Marxist Mask and Romantic Face: Lukács on Thomas Mann," in *Alasdair MacIntyre's Engagement with Marxism: Selected Writings 1953–1974*, ed. Paul Blackledge and Neil Davidson (Boston: Brill, 2008), 321.

[24] MacIntyre, "Marxist Mask and Romantic Face: Lukács on Thomas Mann," 327. Here one sees a precursor to his critiques of the postmodern subject discussed in the next section.

[25] Michel Foucault, "What Is Enlightenment?," in *Ethics, Subjectivity, and Truth*, ed. Paul Rabinow (New York: New Press, 1998), 315–16.

[26] See Badiou, *Saint Paul*.

> originator ... He either does not recognize it as his own or believes ... to have originally seen it in its present form.[27]

Fleck's idea of the thought collective shows how an original insight is brought to a more complete form through processes of dialogue in which the participants can never be sure who contributed what to the idea. In this way, it is the larger scientific community that actually creates a theory, so the moment of insight is only the initiating event of a communal process of intellectual construction. Similarly, religious insights are brought to greater clarity in the community of the church.

Moreover, the moment of insight occurs anew for every generation that is educated into a scientific paradigm. "The learner, like the discoverer, must believe before he can know."[28] For every student, the process of education brings about a moment of perceptual transformation. While perhaps not as exciting as an individual discovery, it is still a conversion of vision and prepares the way for new insights. The church does something similar by mediating the deposit of revelation to each new generation and forming every believer to have her own experience of God.

Practices and Subjectivity

The challenge to one's existing forms of thought that occurs in the event of truth leads to the conversion of the subject, which provides a fertile area for conversation with Christian ethics. Yet many ethicists criticize precisely this instability found in the postmodern subject.[29] Critics like MacIntyre are suspicious of these Nietzschean tendencies, viewing them as an extended argument that one can cast off and replace one's subjectivity and identity like a mask.[30] Instead of integrating oneself through the narrative of a tradition, one jumps from identity to identity. Such a position not only is unsustainable, but also undermines academic conversation, since one cannot take responsibility for one's prior statements if

[27] Ludwik Fleck, *Genesis and Development of a Scientific Fact* (Chicago: University of Chicago Press, 1979), 42–3.

[28] Polanyi, *Personal Knowledge*, 208.

[29] At certain points, Foucault seems to deny the continuity of the subject or at least to desire its discontinuity. "I am no doubt not the only one who writes in order to have no face. Do not ask who I am and do not ask me to remain the same: leave it to our bureaucrats and our police to see that our papers are in order." In Michel Foucault, *The Archaeology of Knowledge* (New York: Pantheon Books, 1972), 17.

[30] Alasdair MacIntyre, *Three Rival Versions of Moral Enquiry: Encyclopaedia, Genealogy, and Tradition* (Notre Dame, IN: University of Notre Dame Press, 1990), 47.

one is always repudiating one's previous identity. There is no way for a discussion to move forward without responsibility.

A first response to these concerns suggests that it is proper that social theory does not seek a uniform, universal model of the subject, instead opting for a more formal definition of the subject. The subject is the self's relation to the self, a relationship that takes different forms in different cultures and philosophies.[31] For a Platonic subject, the relationship to the self takes the form of remembering what one really is, that one is like to the divine. For the Stoic subject, the relationship is one of actively transforming oneself to embody the law governing the universe so that individual reason corresponds with divine reason. The Cartesian subject relates to the self in the form of a method that allows one to advance to certainty. For the neoliberal subject, one relates to oneself as an entrepreneur, a risk-taker who learns various skills of self-branding and self-presentation. Thus, the subject is of central importance, but its form is not a historical constant, and it is crucially important to pay attention to its historical transformations.

A second response is to see this focus on discontinuity as a positive attribute in ethics because the most important question about the subject is that of the possibility of conversion.[32] Can one change the form of subjectivity in which one has been raised? Is the individual forced to remain an entrepreneurial subject or can he become something else? In this model, the transformation of subjectivity occurs only through the attraction of a truth different from the system of truth that supports predominant forms of power. Philosophy attracts one to undertake a different form of life through the critique of prevailing social practices. This conversion does not occur through a brief effort of the will, but requires the long effort of *askesis* that, given the effects of power, may never be fully successful. For my purposes, faith in Christian truth transforms one and leads to continual cooperation in this work of transformation through liturgical engagement, spiritual exercises, and charitable works. Conversion transforms the subject.

In Christianity, this transformation marks a discontinuity in the subject, with the Christian dying to the old self and living anew in Christ.[33]

[31] Foucault, *The Use of Pleasure*, 6; Michel Foucault, *The Hermeneutics of the Subject: Lectures at the Collège de France, 1981–82*, trans. Burchell (New York: Palgrave Macmillan, 2005), 254–8.

[32] Michel Foucault, *Du Gouvernement des Vivants: Cours au Collège de France (1979–1980)*, ed. François Ewald, Alessandro Fontana, and Michel Senellart (Paris: Seuil, 2012), 156; Foucault, *The Hermeneutics of the Subject*, 208–17.

[33] Rom 6:1–11.

Through this conversion, the continuing gift of grace, and, most importantly, Christ's sacrificial death, the Christian is no longer held responsible for his sins. In Christ, the Christian becomes a new creation.[34] Different ethics inspired by Christianity describe this transformation differently. For Aquinas, even if the faculties remain continuous, the infusion of supernatural virtues oriented toward God through grace transforms the individual's habits.[35] For Kant, the individual is transformed in the foundation of his will, changing his fundamental maxim from one directed to empirical desires to one aimed at the moral law. In this way, the self becomes "morally a different human being," which raises difficulties for the question of moral responsibility.[36] Similarly, Badiou develops an interpretation of Paul's writings that sees salvation as the rejection of a prior way of life because of an event, the event of the Resurrection.[37] These moral theories that grow out of the Christian tradition tend to mark a similar discontinuity of the subject.

Because of this consistent theme of death to the old self and rebirth, an insistence on continuity disregards important underlying themes and problems in Christian understandings of the subject. There are ways that responsibility for one's wrong actions before one acceded to the truth of faith does continue under this new form of subjectivity, and the stratified remnants that result from the prior subjectivation remain. In many analyses of Christian ethics, these must be removed, giving an emphasis on mortification and penance for one's previous life. For example, Aquinas argues that even though the newly baptized receives infused virtues, the tracks and orientations of old sinful habits remain in the faculties, and thus virtuous action can seem difficult.[38] He recommends the *askesis* of the evangelical counsels to help transfer our final end from worldly goods to God.[39] Augustine describes the renewal of the image of God that occurs in the Christian life as requiring a long work of transferring our loves

[34] 2 Cor 5:17.

[35] *Summa Theologica* I.51, 63, 64. My interpretation of Aquinas follows A. Knobel, "Can Aquinas's Infused and Acquired Virtues Coexist in the Christian Life?," *Studies in Christian Ethics* 23, no. 4 (2010): 381–96; William Mattison III, "Can Christians Possess the Acquired Cardinal Virtues?," *Theological Studies* 72, no. 3 (2011): 558–85.

[36] Immanuel Kant, *Religion within the Bounds of Bare Reason* (Indianapolis, IN: Hackett, 2009), 74/83.

[37] Badiou, *Saint Paul*, chap. 5.

[38] For an analysis of this problem in Aquinas, see John Harvey, "The Nature of the Infused Moral Virtues," *Proceedings of the Tenth Annual Convention of the Catholic Theological Society of America* 10 (1955): 172–221; Michael Sherwin, "Infused Virtue and the Effects of Acquired Vice: A Test Case for the Thomistic Theory of Infused Cardinal Virtues," *The Thomist* 73, no. 1 (2009): 29–52.

[39] *Summa theologiae* I–II 108.4.

even after baptism.[40] In the perspective developed here, meditation and exercises are necessary to retrain our perception of the world toward a Christian way of seeing and away from reductionist and entrepreneurial ways of seeing the world. In Christian ethics, responsibility must be taken for the relationships harmed by our prior style of life. Yet it is difficult to see why one would be required to take responsibility for statements and discussions prior to embracing a new system of truth, other than by saying that they were wrong and that now one sees the world in a different way. The discontinuity of the subject in this model of practice is far from a liability for a conversation with Christian ethics but is in fact a strength because it helps to inform understandings of conversion.

Truth and Ethics

Despite these strengths, there are also significant defects in any ethic deriving from a formal social theory from the standpoint of moral theology because it will generally lack two elements that are central to any substantive ethics. An ethic needs both (1) a system of truth and (2) an ethical *telos* in the form of something like an ethical exemplar or an ideal form of life. Foucault, for one, fails to provide either in his writings because of his methodological and philosophical commitments, which prevent him from attempting to create a substantive, universal ethic.

Christian ethics, in contrast, provides both truth and an exemplar in Jesus Christ as *Logos*. Creation through the *Logos* ensures that the created order, including human nature, is rationally comprehensible. Therefore, a *Logos* theology undergirds the possibility of the success of any human inquiry, including inquiry into ethics. Because this creative *Logos* was incarnate, Jesus Christ also offers an ethical exemplar to show the form of life by which Christians can attain their ethical *telos*, imaging God. Through a *Logos* theology, Christianity possesses a substantive ethic that a formal system of practice theory lacks.

To address the first element, every moral system is related to a set of substantive truth claims validated by a standard set of procedures.[41] Just as power and knowledge work together to shape subjects, the ethical agent works to transform a *logos* into an *ethos*. One can see these links in

[40] Saint Augustine, *The Trinity*, ed. John Rotelle, trans. Edmund Hill (New York: New City Press, 2012), bk. 14.17.

[41] Michel Foucault, *The Courage of Truth: The Government of Self and Others II: Lectures at the Collège de France 1983–1984*, trans. Graham Burchell (New York: Palgrave Macmillan, 2011), 8–9.

Hellenistic ethics. For the Stoics, the three parts of philosophy – physics, logic, and ethics – were densely interconnected.[42] Similarly, Epicurean physics was necessary to provide the foundation of their ethics by driving away fear of the gods. Even the Cynics, who seem to hold the fewest doctrinal commitments among ancient philosophies and thus make philosophy almost a pure form of life, still had a minimal theory of nature.[43] Without a substantive vision of the world, one cannot have a substantive ethics. For these reasons, a change in ethics and politics must be tied to a changed system of truth.

Foucault and other postmodern theorists do not provide a system of truth that could form the foundation for an ethic. The historical nature of truth claims about the natural world that modern scholarship reveals undermines a philosophy like Stoicism that requires a fairly complete system of the world for ethics. Moreover, Foucault rejects the idea that this history of truth claims is progressing toward any kind of final model of the world, aside from a few disciplines in the hard sciences.[44] While the eschatological inbreaking of the event of truth plays a central role in Foucault's thought, he lacks the hope for an eschatological fulfillment of truth. This lack poses a serious problem for his system because, as Northcott argues in regard to the environmental movement, "without that transcendent reference, . . . protest is still at risk of cynicism and boredom, despondency and hopelessness."[45]

The historical nature of knowledge and its construction through scientific practices poses a problem for Christianity because Christianity promises a grasp of truth. For Christians, Jesus is the way, the truth, and the life.[46] As the second person of the Trinity, he is the *Logos* through whom the world is created. This commitment to the idea that Creation occurs through a rational *Logos* provides an assurance that the world is conceivable by the human mind that is formed in the image of God and is thus rational. It is this assurance that underwrites the possibility of the pursuit of science.

Jesus as incarnate *Logos* creates a different paradigm for the historicity of truth, however, because it is through the Incarnation that God as *Logos*

[42] A. A. Long and D. N. Sedley, *The Hellenistic Philosophers* (New York: Cambridge University Press, 1987), 158–62.

[43] Foucault, *The Courage of Truth*, 189, 263.

[44] Prado, *Searle and Foucault on Truth*, 132.

[45] Michael S. Northcott, *The Environment and Christian Ethics* (New York: Cambridge University Press, 1996), 39.

[46] Jn 14:6.

enters history. The truths arising from the *Logos*, namely, nature (cosmological and evolutionary history), humanity (history proper), and the relation between God and world (salvation history), can be understood as historical, so the human relationship to truth becomes historical. John Henry Newman and MacIntyre have shown that even theological doctrines develop from the original givens of faith.[47] Thus, there is no need to be concerned if knowledge in science or philosophy also shows historical development.

While the historical nature of knowledge may be reconciled with Christian notions of truth if this history is seen in terms of continual or dialectical progress, it is more difficult to see how the history of ruptures, discontinuities, conceptual transformations, and outright misinterpretations that the French tradition of the philosophy of science reveals can be similarly reconciled to Christianity. One fears that there is merely, in Kuhn's terms, a progress from the origins, but not a progress toward anything like a final truth.[48]

It is here that MacIntyre provides a solution. His rationality of traditions is a model by which truth can be both historical, in the way that Foucault's history of science portrays, and pointed toward an eschatological fulfillment.[49] Progress can be seen only from within one of the many traditions of knowledge, and it is only by developing a narrative within such a particular tradition that one can actually see continuity in a seemingly contingent and arbitrary history.[50] Crises and ruptures are essential because it is through these that different traditions encounter each other, that a tradition's weaknesses are truly engaged, that advancement actually occurs.[51] The rupture becomes a moment of progress rather than a mere transformation. However, such a narrative of the development toward truth can be told only from within a certain community and tradition.

Therefore, a tradition of truth will be tied to a community that upholds it through its narrative and practices. For this book, these communities are

47 John Henry Newman, *An Essay on Development of Christian Doctrine* (Notre Dame, IN: University of Notre Dame Press, 1989); MacIntyre, *Whose Justice? Which Rationality?*, 353–4.

48 Thomas Kuhn, *The Structure of Scientific Revolutions*, 3rd edition (Chicago: University of Chicago Press, 1996), 170.

49 MacIntyre, *Whose Justice? Which Rationality?*, 349–69; MacIntyre, "Epistemological Crises and Dramatic Narrative"; Alasdair MacIntyre, "First Principles, Final Ends, and Contemporary Philosophical Issues," in *The Tasks of Philosophy: Selected Essays*, vol. 1 (New York: Cambridge University Press, 2006), 143–78.

50 MacIntyre, "Epistemological Crises and Dramatic Narrative," 10 and 18.

51 MacIntyre, *Whose Justice? Which Rationality?*, 366.

the Church and the community of science. Science is a community that forms its members in a particular narrative of its history. This narrative is governed by an ideal of progress that, as historians of science have shown, is not apparent from all points of view but that is revealed as a regulative ideal of the ultimate unity of science as an accurate description of nature.[52] This goal of the unity of science has often had a reductionist flavor, as in Descartes or the Vienna Circle, for example, seeking to reduce all explanation to mechanism or the simplest observational components. But this reductionist style is not necessary for one to aim at the unification of science, as evidenced by those, like Pierre Duhem, who imagine a final science shaped by Aristotelian metaphysics.[53] The history of science can be read as progress only within certain traditional narratives, and the Christian narrative of progress in science and theology relies on a belief in the rational ordering of the universe, a belief supplied by Christian conceptions of Christ as *Logos*.

Even for science, this idea of truth is only an eschatological concept. Almost all scientists admit that some contemporary scientific theories will need to be completely altered in the future. We do not know which parts of our knowledge are wrong, which concepts out of place, or how things need to be reinterpreted. Honest doubts about specific aspects of contemporary science that arise in scientists or in the public need not become nihilistic skepticism as long as such doubts are tied to a belief that science is actually attempting to describe reality and that that description is attainable as a limit case. If such a commitment to truth relies on a particular narrative in science, then that narrative is even more important for philosophical, theological, and ethical analysis. Because of this, for Christians, a history of rupture must be replaced by a rationality of traditions.

While a moral system is always linked to a system of truth, it also requires an ethical *telos*, the mode of being as a subject that one seeks to attain through ethics, a mode that can be seen in the moral exemplars one emulates. Given the nature of his perspectivism and his distaste for directing others through his work, it is unclear that Foucault could provide a normative *telos*. Instead, he gives descriptive tools that can be used to

[52] MacIntyre, "First Principles, Final Ends, and Contemporary Philosophical Issues," 149, 165–6.

[53] Pierre Duhem, "Physics of a Believer," in *The Aim and Structure of Physical Theory* (Princeton, NJ: Princeton University Press, 1954), 273–311.

analyze a number of ethical frameworks, be they classical, Hellenistic, Christian, Rabbinic, or Islamic.[54]

Since this model of practices of truth is a formal system created partially through an analysis of and abstraction from Christian ethics, though, it is not surprising that it can easily be filled with the substance of Christian ethics. For Christianity, Jesus Christ exemplifies the ethical *telos*. As Incarnate *Logos*, he most truly embodied the *logos* of human created nature in a form of life. Further, as a rational creature made to the image of God, the human *telos* is a likeness of God. As second person of the Trinity, Jesus is both the image of God and fully God, and the Christian form of life is one of imitation of Christ. Instead of merely differences of formal similarities, i.e., substituting Jesus as exemplar for Socrates, Jesus' role goes beyond what a secular moral philosophy could envisage because he is not only a model to which we aspire but also the Mediator of the grace that is the means to attain the ideal form of subjectivity. The importance of this fact will be seen in the next section discussing the limited hopes the Stoics had for their ethics. It is by providing this Mediator that Christianity goes beyond such forms of philosophical ethics.

The *Imago Dei* and the Ethics of Practice

Rather than ending on this note of where a substantive Christian ethics perfects the formal ethics of social theory, I will conclude by further investigating what lessons MacIntyre's Aristotelian and Foucault's Stoic ethics have for more traditional Christian understandings of theological anthropology. This exercise helps to show the broader relevance and applicability of this project for Christian ethics and anthropology. I will take the example of the theological anthropology found in the Augustinian tradition of the *imago dei* because it is one of the dominant understandings of the human person in the Western tradition. Beyond its theological importance, it also has many points of theoretical contact with the work discussed here because this tradition recognizes the same profound

[54] For applications of Foucault's ethics to Judaism and Islam, see Jonathan Schofer, *The Making of a Sage: A Study in Rabbinic Ethics* (Madison: University of Wisconsin Press, 2005); Mahmood, *Politics of Piety*. In his late writings, Foucault gives many suggestions about what an ethical *telos* could be in modern life. Each person could choose his own, creating new forms of life. Alternatively, one could emulate the modern artist, treating life as an aesthetic project. Or one could embrace the ideal of the philosophical parrhesiast, speaking risky truths to the powerful. Though some have tried to reconstruct a substantive ethic out of these hints, all of these attempts are rather problematic.

connection between knowledge and subjectivity. Humans are rational creatures who know and love themselves and God through the intellect and will, which are in the image of the activities of the *Logos* and Spirit. Knowledge is tied to affection and action because one's knowledge of self and God changes who one is. Conversely, false images of the self deform one's desires and subjectivity. Reformation comes through gaining a better grasp of who one is as made in the image of God. Rather than looking at later thinkers in this tradition of Augustinian anthropology like Aquinas, Calvin, or Luther, I will focus on Augustine's discussion in *De trinitate* both because it is a foundational text of the tradition and because it is more closely linked to the style of thought of ancient philosophy, although not determined by it. While this section shows the congruence of the formal conception of subjectivity developed here with the more substantive analysis of Augustine, the focus of Foucault and MacIntyre on practice can help to correct problems that arise later in the Augustinian tradition from its emphasis on the problem of desire. This section explores how the model of ethics examined in this book may alter aspects of Christian anthropology, while also showing how theological anthropology works to fill substantive gaps in formal ethical systems.

Misperceiving the Self

In Augustine's *De trinitate*, there is a deep relation between the knowledge of one's self and mind and one's knowledge of the Triune God since humans are made in the image of God.[55] While faith in Revelation gives correct rules for thinking and speaking about God, the Christian should go further to actually seek understanding of that professed faith.[56] One way to

[55] E.g., "Through this image, which we are, we might see Him by whom we have been made in some manner or other, as through a mirror"; Augustine, *The Trinity*, bk. 15.8.14. My discussion of *De trinitate* is influenced by Rowan Williams, "Sapientia and the Trinity: Reflections on De Trinitate," in *Collectanea Augustiniana: Melanges T. J. Van Bavel*, ed. Bernard Bruning, Mathijs Lamberigts, and J. van Houtem (Louvain: Leuven University Press, 1990), 317–32; John Cavadini, "The Structure and Intention of Augustine's De Trinitate," *Augustinian Studies* 23 (1992): 103–23; Michel Rene Barnes, "Augustine in Contemporary Trinitarian Theology," *Theological Studies* 56 (1995): 237–50; Oliver O'Donovan, *The Problem of Self-Love in St. Augustine* (Eugene, OR: Wipf and Stock, 2006); John Cavadini, "The Darkest Enigma: Reconsidering the Self in Augustine's Thought," *Augustinian Studies* 38, no. 1 (2007): 119–32; Luigi Gioia, *The Theological Epistemology of Augustine's De Trinitate* (Oxford: Oxford University Press, 2009); Paul R. Kolbet, *Augustine and the Cure of Souls: Revising a Classical Ideal* (Notre Dame, IN: University of Notre Dame Press, 2009); Lewis Ayres, *Augustine and the Trinity* (Cambridge: Cambridge University Press, 2010); Lydia Schumacher, *Divine Illumination: The History and Future of Augustine's Theory of Knowledge* (Malden, MA: Wiley-Blackwell, 2011).

[56] Augustine, *The Trinity*, bk. 9.1.

grow in this understanding is through the spiritual exercises by which one see likenesses of the Trinity in the knowledge and love of self.[57] Such exercises are always imperfect and are never enough to give the vision of God granted in the afterlife, but they are a start and can serve to purify one's thoughts.

This strategy has classical precedents going back at least to Plato's *Alcibiades I*, in which Alcibiades was to obtain knowledge of self by meditating on the divine, much as the eye sees itself in a mirror.[58] Such a strategy connecting the soul and God continues in the Stoics and Neo-Platonists. The Stoics described the perfect rationality of virtue in terms of an image or likeness to God because God is Reason, *Logos*.[59] As Musonius Rufus says, "So man, as the image of [God] when living in accord with nature, should be thought of as being like him."[60] Similarly, according to Cicero, "Virtue exists in man and God alike, but in no other creature besides; . . . there is therefore a likeness between man and God."[61] This Stoic analysis reaches back at least to the second head of the Stoa, Cleanthes, who, in one reading of a disputed line of his *Hymn to Zeus*, says, "For we have our origin in you, bearing a likeness to God, we alone of all that live and move as mortal creatures on earth."[62] It takes rational work to develop a coherent vision of oneself as a part of the rational cosmic order, but once one gains this coherent conception of self as like to the divine, then one becomes a sage.

The fundamental problem with such simple attempts by the self to image the divine was noted in both Augustine and these classical precedents: a Fall away from the divine into the particular or material. For Augustine, this Fall deformed the image of God in humans. As the Stoics also perceived the problem, humans focus on the supposed good of the individual's particular part of the universe, thereby misunderstanding the connection between one's own good and the good of the whole universe, the good willed by that universe's Creator. "For when the soul loves its

[57] See my discussion in Chapter 5 of Gilles Emery, *Trinity, Church, and the Human Person: Thomistic Essays* (Naples, FL: Sapientia Press, 2007).

[58] Plato, "Alcibiades I," in *Complete Works*, ed. John Cooper, trans. D. S. Hutchinson (Indianapolis: Hackett, 1997), 132d–133c.

[59] This aspect of Stoicism is part of a broader theme in ancient philosophy arguing that the aim of ethics is assimilation to God. See Geurt Hendrik van Kooten, *Paul's Anthropology in Context: The Image of God, Assimilation to God, and Tripartite Man in Ancient Judaism, Ancient Philosophy and Early Christianity* (Tübingen: Mohr Siebeck, 2008), 124–81.

[60] Fragment 17 in Johan Carl Thom, *Cleanthes' Hymn to Zeus: Text, Translation, and Commentary* (Tubingen: Mohr Siebeck, 2005), 61.

[61] *De legibus* 1.25 cited in Thom, *Cleanthes' Hymn to Zeus*, 61.

[62] Thom, *Cleanthes' Hymn to Zeus*, 40.

own power, it slips from the common whole to its own particular part."[63] Moreover, these fallen desires lead to a false image of the self. Taking material objects, what one might call natural goods or the preferred indifferents of the Stoics, for overriding goods, the individual ties his will to them, conforming his intellectual capacities to them rather than to the divine *Logos*. "The mind errs when it binds itself to these images [of corporeal things] with a love so strong as even to regard itself as something of this kind. And thus it is made like them to some extent, not in reality but in its thoughts."[64] One sees one's soul as similarly material and becomes utterly mistaken about what one is.

While Augustine speaks about this misperception of the self resulting from wrong desire, MacIntyre and Foucault provide a perhaps more helpful path by showing how practices warp the image of the self. When one thinks of the image within as disfigured, one should think not about desires enmeshing the self with material things and thereby giving false images of the self. Instead, one ought to examine practices that generate images of the self that prevent the self from coming to the likeness of God.

Such false images can be better termed idols. A turn to the problem of idolatry shows the relevance of biblical analyses for these problems of the connection between knowledge of the self and the self's moral formation. Turning to these scriptural sources reveals the possible connections between idolatry and technological practice.[65] For example, in Isaiah and the Wisdom of Solomon, idolatry is intimately tied to craft practice, to *technai*.[66] The woodcutter, the potter, and the metal worker deploy their skills in response to idleness, weakness, or mechanisms of power to create an object for worship. "This became a hidden trap for humankind, because people, in bondage to misfortune or to royal authority, bestowed on objects of stone or wood the name that ought not to be shared."[67] One could add to these older crafts the dangerous idols that emerge from the *techne* of modern biomedical science. The worship of idols results from a false view of the world, reinforcing systems of power and playing on human weakness. Importantly, engagement with idolatry shapes the self. In creating idols and trusting in them, one forms oneself in their image rather than God's:

[63] Augustine, *The Trinity*, bk. 12.9.
[64] Augustine, *The Trinity*, bk. 10.6.
[65] For a different interpretation of the ties between technology and idolatry, see Brian Brock, *Christian Ethics in a Technological Age* (Grand Rapids, MI: Eerdmans, 2010), 225–34.
[66] Is 44:9–20; Wis 13:10–15:17.
[67] Wis 14:21.

> The idols of the nations are silver and gold, the work of human hands. They have mouths, but they do not speak; they have eyes, but they do not see; they have ears, but they do not hear, and there is no breath in their mouths. Those who make them and all who trust them shall become like them.[68]

For an example of the application of such concepts of idolatry to contemporary neoliberal society, one need only look to the "idolatry of the market." Such idolatry results from deforming the initially good aim of creating markets for ease of exchange. Almost everyone recognizes the market's allocative efficiency, but, as discussed in Chapter 2, some theorists have elevated it to a perfect information processor that should be universally deployed. It is in this false knowledge that idolatry threatens. One's *logos* becomes that of a market rationality. As markets are deployed in more areas of life, and as individuals become increasingly embedded in them, people begin to see more and more of the creatures of the world as commodifiable and humans as primarily market actors. Thus, individuals increasingly turn objects in the world into commodities and develop themselves in terms of human capital. In an accelerating, circular process, markets form worldviews and worldviews transform the world. Eventually, social policies are put in place that form individuals into entrepreneurial subjects. It is here that the image of God is defaced. The *telos* of forming oneself into the likeness of God is replaced through these social processes by the *telos* of the entrepreneurial subject. To be capable of seeking the likeness of God, one must transform the image of the entrepreneurial subject within oneself and detach oneself from the rationality and practices that form it.

A similar process occurs with a reductionist worldview, one much closer to Augustine's description of the soul's relation to material objects. The biologist's reductionist conception of the body is reciprocally linked to reductionist experimental manipulations of bodies. As the scientist becomes embedded in a reductionist worldview through daily practice, rationality is reduced to an instrumental rationality. This process causes the disciplines of medicine and public health to treat the body in a mechanistic manner. This development can end in genetic or neural reductionist accounts of the human. These accounts liken the soul to biochemical matter due to intellectual and practical attachment to material things. This vision can also be deployed in authoritarian or neoliberal

[68] Ps 135:15–18. For a thorough analysis of the theme of formation in the image of idols, of becoming like the works of our hands, but without my focus on *techne* or practice, see G. K. Beale, *We Become What We Worship: A Biblical Theology of Idolatry* (Downers Grove, IL: IVP Academic, 2008).

eugenic attempts to shape the person and the social body through the elimination or shaping of the genetics of the population. The subject comes to be seen in a mechanistic manner.

This analysis can deflect criticisms of Augustine's *imago dei* as too private and individualist,[69] because it reveals the social formation of problematic images of the self and the role that a personal conversion of that image can play in reforming social practice. The criticism of *De trinitate*'s individualism misses many of its aspects: the tie of the image to justice and love, the larger context of Augustine's criticism of Roman imperial power, or the turn to Christ as involving service to others.[70] Still, many people would prefer an analysis of the *imago dei* that takes advantage of the knowledge of social structures and uses the *imago dei* to confront structures of injustice. The analysis of subjectivation in social theory provides these resources. It shows how the self is malformed through the processes of subjectivation, describing how these defective images of the subject as entrepreneur or mechanism perpetuate problematic relations of power. From a Christian perspective, false rationalities lead to problematic subjectivities. A deformed *logos* is embodied in a sinful *ethos.* More importantly, the conversion and renovation of the individual subject in that subject's conception of the self is directly tied to the capacity to challenge contemporary technologies of power.[71] We need to alter social practice in order to alter our forms of subjectivity. By transforming the self, one is enabled to help transform others and change the systems that prevent their flourishing. Thus, we can connect Augustine's meditative practice on the Trinity to the transformation of general social practice.

Transforming the Vision of the Self

The question is how to change the vision of the self. In a positive vein, Augustine leads the reader through spiritual exercises and directs him to practical action. He approves of Cicero's advice in the *Hortensius* to apply oneself to investigating with reason as well as avoiding human vice and delusion, although with the proviso that one cannot rely on reason alone.[72] One grows in the love of God through such contemplative exercises and

[69] E.g., Catherine LaCugna, *God for Us: The Trinity and Christian Life* (San Francisco, CA: Harper, 1993).

[70] Cavadini, "The Darkest Enigma: Reconsidering the Self in Augustine's Thought," 128.

[71] Foucault, *The Hermeneutics of the Subject*, 252.

[72] Augustine, *The Trinity*, bk. 14.19.

through the imitation of Christ in love and justice toward others.[73] "By our striving we imitate Him who remains and follow Him who stands; when we walk in Him, we tend towards Him, because by His humility He has been made a road for us in time, in order that by His divinity He might be for us a mansion in eternity."[74] Augustine's system is compatible with the practices of the self described here.

Negatively, Christians must disengage their desires from the material objects that now shape them:

> This renewal, of course is not brought about in the one moment of the conversion itself ... Whoever ... is being renewed in the knowledge of God, and in justice and holiness of truth, by making progress day by day, transfers his love from temporal to eternal things ... and constantly endeavors to restrain and to lessen the desire for the former, and to bind himself by love to the latter. But he does so in proportion to the divine help that he receives.[75]

However, the negative task of purifying one's self-image can take dangerous turns in the later Augustinian tradition, especially in the hyper-Augustinian line of that tradition.[76] These dangers are partly rooted in Augustine's focus on desire, which, while not leading to issues in the work of Augustine himself or even in many of his successors, can give rise to two problems. First, much as in the Aristotelian tradition, once desire is malformed, it is unclear how to reform it. One may trust in God's grace for reform, an assumption necessary for any practical Christian ethic, but it is not obvious what steps the individual as agent can take to cooperate with that grace. If one seeks to cooperate with God's grace in one's formation, the only clear path in the Augustinian tradition is mortification of desire.

Yet this option leads to an even deeper problem with a focus on desire.[77] Beyond the problem of mortification tempting one to a mind–body dualism that has always threatened Christianity, a dualism that would cause the Christian to despise the body that the Incarnation teaches her to love, a dualism all too complementary to reductionist and entrepreneurial reworkings of the body, there is the question of what the actual desire is

[73] Augustine, *The Trinity*, bk. 8.8.
[74] Augustine, *The Trinity*, bk. 7.3.
[75] Augustine, *The Trinity*, bk. 14.17.
[76] For discussions and analysis of hyper-Augustinianism, see Charles Taylor, *Sources of the Self: The Making of the Modern Identity* (Cambridge University Press, 1992), 246–7; Jennifer A. Herdt, *Putting on Virtue: The Legacy of the Splendid Vices* (Chicago: University of Chicago Press, 2012).
[77] Much of my discussion here draws from Herdt, *Putting on Virtue*.

that drives mortification.[78] Is it truly to form oneself in the image of God in order to serve the Lord? Or is it actually a desire to glorify oneself, to show strength and self-will, and to gain honor in the sight of the world? The truth of desire is not obvious. It does not lie close to the surface of consciousness in the Western Christian tradition. The spiritual director or confessor must dig through the layers of false consciousness to find true intentions and aims, forcing the penitent to turn desire into discourse.[79] The source of thoughts must be discovered – God or the devil – and the self must be decoded.[80] Spiritual exercises become worse than useless in the later hyper-Augustinian tradition. As Jennifer Herdt notes, "For Luther . . . there is no route from 'external' practices to fundamental inner transformation . . . The spiritual exercises . . . will lead to hypocrisy and false piety."[81] In this hyper-Augustinian development of the tradition, even virtues become vices resulting from pride or hypocrisy, and human agency itself seems undermined.

It is for this reason that Foucault criticizes the emphasis on desire. Contemporary theologians have good reason to be concerned as well because this problem space can easily be inverted, so that true desire is no longer something to be deciphered so that it can be altered but rather becomes something to be sought so that it can be celebrated. In this view, moral tradition comes to be seen as repressive, as something to be cast off in order to achieve the happiness brought by true desire. Even when given a communitarian gloss, such an ethics of authenticity can lead to an anxious search for one's true individuality, one's authentic self, one's actual desire, none of which may exist. Moreover, it leaves one open to be governed by those others who will help to excavate the self, to the advertisers whose products will express who you truly are, to the business manuals that teach how to develop one's brand, to the rebels who aid in the rejection of authority while at the same time commodifying dissent, to the pharmaceutical companies and surgeons who can help suppress those elements of the self that do not fit the vision of the ever youthful and happy true self.

In contrast, the Christian is called to live in Jesus, a call leading not to a true subject to be deciphered under layers of accretion, but to a task for the Christian to complete with God's help. One is called to a likeness of God

[78] Foucault, *Security, Territory, Population*, 205–7; Foucault, *Du Gouvernement des Vivants*, 286.
[79] Michel Foucault, *The History of Sexuality*, vol. 1 (New York: Pantheon Books, 1978), 21.
[80] Foucault, *Du Gouvernement des Vivants*, 291–305; Foucault, *The Hermeneutics of the Subject*, 256.
[81] Herdt, *Putting on Virtue*, 174.

that is not yet present in the mere formality of the image of self-knowledge and self-love. There is no true self waiting for discovery, but a death to the old self. The decipherment of desire demanded by the contemporary project of the self is an infinite task since the depths of the soul are known only to God.[82] While loves and desires are critical to reformation, these are domains in which the agent cannot directly operate. It is only God who can work on them through the gift of charity, and one can only trust that God is active there.[83]

What can the individual do then? MacIntyre and Foucault direct our attention to practice. While MacIntyre helpfully points to the role of traditioned social practices, Foucault points toward personal practice, recommending a twofold strategy. One needs to engage in different forms of truth and practices of the self.[84] Stoicism provides the most useful kinds of practices of the self for rehabilitating the image of God in the mind. As the Stoics taught, one can operate only on those things immediately under the one's control: how one approaches representations and what practices or exercises one adopts. One can give or deny assent to our representations of the good of an exterior object, thus shaping one's impulses toward it.[85] The individual shapes daily practices and exercises in order to mold her approach to representation and to control her response to temptations. It is on these aspects of the self that the Stoics teach us to act.

Yet Christians have denounced these pagan and postmodern attempts to achieve virtue through self-work as merely expressing the pride of fallen nature. Such reliance on self-work can seem to be Pelagian because it disdains the grace necessary to heal broken selves.[86] It expresses a presumption that we will be able to achieve salvation through our own efforts. These criticisms, however, miss an important aspect of this form of thought. Foucault doubts whether the self-transformation for which he argues is possible, mournfully concluding, "I do not know whether we will ever reach mature adulthood."[87] His work is a call to action,

[82] Augustine, *The Trinity*, bk. 15.7.

[83] Here, I do not analyze the mechanics of God's graced action on our desires, but for an interpretation of Augustine's thought on this topic interpreted in a Stoic vein, see Sarah Catherine Byers, *Perception, Sensibility, and Moral Motivation in Augustine: A Stoic-Platonic Synthesis* (New York: Cambridge University Press, 2013), 179–207.

[84] Foucault, "The Ethics of the Concern for Self as a Practice of Freedom," 295, 299.

[85] Epictetus, *Discourses*, Bks. 1–2, trans. W. A. Oldfather (Cambridge, MA: Harvard University Press, 1998), I.1.

[86] For a discussion of the dangers of presumption and Pelagianism in other contemporary strands of Christian ethics, see David Elliot, *Hope and Christian Ethics* (New York: Cambridge University Press, 2017), 112–20.

[87] Foucault, "What Is Enlightenment?," 318.

not a prediction of victory. The Stoics also doubted that success was possible. They were as skeptical of the possibility of natural happiness as any Christian writer. Most Stoics thought that at most perhaps one, maybe two, sages had ever existed.[88] Almost everyone saw Socrates as a sage, and Seneca embraced Cato as well, but otherwise, it seemed nearly impossible that a person could fully heal his reason. This conclusion is even more striking when one considers that, for the Stoics, virtue is all or nothing. If one is not a sage, then all of one's dispositions and actions are vicious. There is no need to look to Augustine for the concept of the splendid vices, the claim that the seeming virtues of most pagans were really vices, because it was essential to the Stoic system that barely anyone was capable of forming true virtue. Most people can only try to exercise the self in the hope of progressing toward the moment of the transformation of the self through the truth of wisdom.

Christians can approve of this skepticism. The person needs Christ as exemplar but also as Mediator of the grace that heals and transforms the self. Human efforts alone are not enough to attain a true vision of the self. The only way to banish the idols that impress themselves on the subject is with the divine aid provided both directly and as an example. As the *Alcibiades I* suggests, humans can know themselves only by looking at another exemplar as the eye sees itself in the mirror, and even with such an exemplar the person in his current deformed state still requires aid. It is the Incarnation that provides this aid as well as this true image of the virtuous way of life. Through his death on the Cross, Christ both served as a necessary exemplar of the humility required for humanity to again focus on the good of the whole by following God's will and rescued humans from slavery of sin.[89] Through Christ's action "the pride of man, which more than anything else hinders him from cleaving to God, might be refuted and cured by such great humility on the part of God."[90] Through the Holy Spirit, God continues to support individuals on the path of salvation. However, an emphasis on the importance of the practices of the self that help to conform one to this image is also essential to contemporary moral theology. Theology can fulfill secular visions of the connections between *askesis*, truth, subjectivity, and salvation by pointing to Christ, who is the way, the truth, and the life.

[88] Long and Sedley, *The Hellenistic Philosophers*, 325, 381, 383.
[89] Augustine, *The Trinity*, bk. 13.16.
[90] Augustine, *The Trinity*, bk. 13.17.

Conclusion

This book has argued that contemporary philosophy and social theory give us important tools for understanding the current predicament of science. Yet secular theorists offer little real hope for change in themselves. Previous chapters have discussed the problems of the Aristotelian model based in habituation through communal practice. On the other hand, the social theoretical, Stoic model found in Foucault offers no future hope, no possibility for an eschatological achievement of truth, and no ethical *telos* as a model. For this reason, the Aristotelian and Stoic models serve as useful supplements to each other. The Stoics as discussed in Foucault's work explain the importance of conversion for virtue ethics, conversion wrought through an inbreaking of truth that shows the distortions of the current knowledge of self and world and of the possibility of a different way of being. It is a conversion brought to fruition through extensive work on the self. MacIntyre's rationality of traditions offers the promise that the truth that converts is not just a chance shift of disciplinary structures. It could be actually a development toward a fuller truth. Both perspectives are necessary for an adequate understanding of moral formation.

In Christian ethics, both of these aspects of hope depend on Christ. It is Christ as *Logos* who assures the rationality of created existence. It is Christ who through his death on the Cross and through whose mediation of grace makes conversion possible. And it is Christ who through His Incarnation and life on earth shows the form of life at which self-work should aim. Through Christ, subjectivity can be transformed, leading to ultimate flourishing. It is through the Christian revelation that the Stoic philosophical system can lead beyond an ultimately unattainable happiness to actual fulfillment.

Conclusion

A little over a hundred years ago, Max Weber gave an enduring description of what it means to pursue a career in science as a vocation. He portrayed it as a difficult pursuit in the context in which he wrote, the pre-Nazi German research university. Even then, it required asceticism, devotion, and an ability to weather repeated disappointments. Even then, academics were under pressure to become prophets or to subordinate their search for truth to political or institutional ends. Yet today, Weber's time seems like a golden era of scientific research, giving rise to fundamental insights into physics, applied chemistry, and engineering that still form the foundation for much of our current understanding of the world. It gave rise to bold experiments in research communities such as the Kaiser Wilhelm Institutes.

In contrast to Weber's age, the structures governing contemporary scientific research create many more pressures against seeing a career in science as a vocation dedicated to pursuing the truth. Researchers are directed to pursue short-term possibilities for commodification, pressured to shape their careers to maximize poorly designed metrics of publications, forced to compete for too few grants, and given neither the time nor support to pursue long-term, risky research ventures that might illuminate more fundamental understandings of the world around us. It is little surprise that these career circumstances incline individuals toward hyping their results, rushing their findings into publication before they are entirely well founded, focusing on the research paths of least resistance, and, in the most extreme cases, committing fraud and other forms of misconduct. The social theories discussed in this book offer a diagnosis of how contemporary practices like those in science deform knowledge of self, ways of seeing, dispositions, and actions. They show the mechanisms through which individuals, especially contemporary scientists, are problematically formed in reductionist and market-based ways of approaching the world. The central impetus of this book, echoed by many other commentators, is to change this system of practice.

The problem lies not only in the structures, though. The real problem is primarily the ideal that motivates many of these structures, the ideal of entrepreneurial science. It is an ideal that distracts one from the pursuit of truth toward the pursuit of profit, taking advantage of the instrumentalist stance toward the world imbued by the daily practice of laboratory research. This ideal, the entrepreneurial character that it shapes, and the reductionist way of viewing the world that it draws on, must be transformed in contemporary research, so that researchers can again develop virtues directed toward the end of the scientific vocation, the pursuit of truth.

As I have argued, the standard approach to virtue theory in the field of Christian ethics, the Aristotelian model, will not successfully address these issues. This model depends on a community with practices that can habituate researchers into the required virtues and character. Such a community does not exist in contemporary science. It is not that there are not good laboratories or good people throughout the research community – there are – and I was lucky enough to experience more than my fair share of them during my research training. Yet these islands of good practice are constrained by the structures of research, minimizing the good that they can do and leaving students to face the competitive forces of the wider field alone once they leave their training ground. Moreover, fewer and fewer researchers have had experience in prior modes of performing science, so the entrepreneurial model is becoming naturalized. The goal of this book is to develop ways to help maintain the scientific ideals of those who have them and to form the character traits that help one to achieve these older scientific ideals in those newly entering the field.

One solution to this problem is found in the tradition of spiritual practices that emanated from Stoicism before becoming adapted into forms of Christian spirituality. In their original Stoic form, one can see a role for these practices that can be obscured by their many other purposes in Christian spirituality. They serve to transform how an individual sees the world, shaping her to see it as an order of meaning and purpose in which virtue has the highest value. Such a conversion and maintenance of vision can help prevent the researcher from falling into the generally instrumentalist worldview supported by the practices of science as well as the short-term opportunism that can be encouraged by a misplaced entrepreneurialism. By seeing the world truly, the individual's dangerous passions are healed, and she is redirected to truth and the service of others. It is necessary to form individuals in these practices and to support them through communities dedicated to the understanding of the world embodied in these practices.

My claim is not that simply because individuals engage in these spiritual practices all the problems of science will be solved. There will still be a need for structural changes to fully reform science, many of which I have discussed in passing. This book is not a denial of the need for structural change. Although there have been some positive developments in areas like patent law and the willingness of senior members of the scientific community to call for overhauls of the research enterprise, these structural changes do not seem to be on the horizon right now. Until they are, these spiritual practices can help maintain the character and ideals of individual scientists. These well-formed specific intellectuals can serve as points of resistance to institutional overreach and exemplars of good science that can inspire the next generation. These individuals can also serve as a core to drive the institutional changes that are needed in science. Individual and institutional conversion are not in opposition to one another but are instead complementary. Even once the structures of science change, which they eventually will, ongoing spiritual practices will still be necessary to maintain virtue. If the history of science demonstrates anything, it shows that new institutional structures will bring new pressures and temptations. Every social order contains possibilities for corruption.

While I encourage all scientists and scientific training programs to take up this model of moral formation, I believe that scientists and universities that are Christian have a special role to play in the efforts to reform the ethics of science. Unlike Weber, Christians do not have to summon a heroic resolve to pursue truth in the face of a meaningless world. They are already committed to an understanding of the world as possessing inherent meaning and purpose given to it by its Creator. They thus must take special care not to succumb to the reductionist vision that scientific practice can sometimes form. Their commitment to truth does not rely solely on the choice of a career in science or the formation through the practices of the scientific community. Christians should have an independent source for their commitment to truth, a commitment to discover the *Logos* working in the world around them. Moreover, they should be committed to forming themselves into the image of this *Logos*. They also have an independent community, the Church, that should support them in living out this vocation, despite the problems that have arisen in this relationship historically. They are not forced to stand alone as a heroic individual as it frequently seems that Weber's researcher does. Thus, a vocation in science can be an ideal way to embody the Christian life.

Finally, Christians already have a long tradition of spiritual practices of the kinds recommended in this book. There is no need for them to develop them de novo or adapt them haphazardly from Stoic texts. There will be a need for some adaptation because the shape of the contemporary life in research, its pressures and rhythms, differ from those forms of life for which many of the great classics of spirituality were written. Still, many writers, like Sertillanges, have made considerable efforts to consider the interrelationship between the Christian and the scholarly life. These resources, properly developed, offer a rich opportunity for developing sets of spiritual practices to shape scientific character. Individual Christians can use these spiritual exercises to fight against the pressures of daily life in the contemporary laboratory. Christian universities can integrate them into their training programs as a way to form students and offer exemplary institutional resistance to problematic trends in contemporary science. In these ways, individuals will truly be able to live out their careers in contemporary science as a Christian vocation.

Bibliography

Alberts, Bruce, Marc Kirschner, Shirley Tilghman, and Harold Varmus. "Rescuing US Biomedical Research from Its Systemic Flaws." *Proceedings of the National Academy of Sciences* 111, no. 16 (2014): 5773–7.

Anderson, Chris. "The End of Theory: The Data Deluge Makes the Scientific Method Obsolete." Wired, June 23, 2008, www.wired.com/2008/06/pb-theory/.

Angell, Marcia. *The Truth about the Drug Companies: How They Deceive Us and What to Do about It.* New York: Random House, 2005.

Angier, Tom. *Techne in Aristotle's Ethics: Crafting the Moral Life.* London: Bloomsbury Academic, 2012.

Annas, Julia. *The Morality of Happiness.* New York: Oxford University Press, 1993.

Annual Report of the Government Chief Scientific Adviser. "Innovation: Managing Risk, Not Avoiding It. Evidence and Case Studies," 2014. www.gov.uk/government/uploads/system/uploads/attachment_data/file/381906/14-1190b-innovation-managing-risk-evidence.pdf.

Anscombe, G. E. M. "Modern Moral Philosophy." *Philosophy* 33, no. 124 (1958): 1–16.

Antonaccio, Maria. "Contemporary Forms of Askesis and the Return of Spiritual Exercises." *Annual of the Society of Christian Ethics* 18 (1998): 69–92.

Aquinas, Thomas. *Summa Theologica*, trans. Dominicans of the English Province. New York: Benziger Bros., 1947.

Arendt, Hannah. *The Human Condition.* Chicago: University of Chicago Press, 1998.

Aristotle. "Nicomachean Ethics." In *Introduction to Aristotle*, ed. Richard McKeon, trans. W. D. Ross, 327–579. New York: Modern Library, 1992.

Aronowitz, Robert. *Risky Medicine: Our Quest to Cure Fear and Uncertainty.* Chicago: University of Chicago Press, 2015.

Artigas, Mariano, Thomas F. Glick, and Rafael Martínez. *Negotiating Darwin: The Vatican Confronts Evolution, 1877–1902.* Baltimore, MD: Johns Hopkins University Press, 2006.

Asad, Talal. *Genealogies of Religion: Discipline and Reasons of Power in Christianity and Islam.* Baltimore, MD: Johns Hopkins University Press, 1993.

Attridge, Harold, ed. *The HarperCollins Study Bible: New Revised Standard Version*. San Francisco, CA: HarperOne, 1989.
Augustine, Saint. *The Trinity*, ed. John Rotelle, trans. Edmund Hill. New York: New City Press, 2012.
Auletta, Ken. "Blood, Simpler." *The New Yorker*, December 15, 2014, www.newyorker.com/magazine/2014/12/15/blood-simpler.
Aviv, Rachel. "A Valuable Reputation." *The New Yorker*, February 10, 2014.
Ayres, Lewis. *Augustine and the Trinity*. Cambridge: Cambridge University Press, 2010.
Badiou, Alain. *Saint Paul: The Foundation of Universalism*. Stanford, CA: Stanford University Press, 2003.
Baer, Drake. "This Korean Lab Has Nearly Perfected Dog Cloning, and That's Just the Start." *Business Insider*, September 12, 2015, www.businessinsider.com/how-woosuk-hwangs-sooam-biotech-mastered-cloning-2015-8.
Baker, Monya. "1,500 Scientists Lift the Lid on Reproducibility." *Nature* 533, no. 7604 (May 26, 2016): 452.
Banner, Michael. *The Ethics of Everyday Life*. New York: Oxford University Press, 2014.
Barnes, Michel Rene. "Augustine in Contemporary Trinitarian Theology." *Theological Studies* 56 (1995): 237–50.
Barrera, Albino. *Economic Compulsion and Christian Ethics*. Cambridge: Cambridge University Press, 2007.
Market Complicity and Christian Ethics. New York: Cambridge University Press, 2014.
Bartelink, G. J. M. "Quelques observations sur parrēsia dans la littérature paléo-chrétienne." In *Graecitas et Latinitas Christianorum Primaeva. Supplementa; Fasciculus 3*, 7–57. Nijmegen: Dekker & Van de Vegt, 1970.
Basken, Paul. "U.S. House Backs New Bid to Require 'National Interest' Certification for NSF Grants." *The Chronicle of Higher Education*, February 11, 2016.
Beale, G. K. *We Become What We Worship: A Biblical Theology of Idolatry*. Downers Grove, IL: IVP Academic, 2008.
Beaulieu, Alain. "Towards a Liberal Utopia: The Connection between Foucault's Reporting on the Iranian Revolution and the Ethical Turn." *Philosophy and Social Criticism* 36, no. 7 (2010): 801–18.
Becker, Gary Stanley. *Human Capital*, 3rd edition. Chicago: University of Chicago Press, 1993.
Begley, C. Glenn, and Lee M. Ellis. "Raise Standards for Preclinical Cancer Research." *Nature* 428 (March 28, 2012): 531–3.
Behrent, Michael. "Foucault and Technology." *History and Technology* 29, no. 1 (2013): 54–104.
"Liberalism without Humanism: Michel Foucault and the Free-Market Creed, 1976–1979." *Modern Intellectual History* 6, no. 3 (2009): 539–68.
Benatouil, Thomas. *Faire Usage: La Pratique du Stoicisme*. Paris: VRIN, 2006.

Berger, Margaret. "Upsetting the Balance between Adverse Interests: The Impact of the Supreme Court's Trilogy on Expert Testimony in Toxic Tort Litigation." *Law and Contemporary Problems* 64 (2001): 289–326.

Bernal, J. D. *The Social Function of Science*. London: Faber & Faber, 2010.

Bhattacharjee, Yudhijit. "Diederik Stapel's Audacious Academic Fraud." *New York Times*, April 26, 2013, sec. Magazine. www.nytimes.com/2013/04/28/magazine/diederik-stapels-audacious-academic-fraud.html.

Biagioli, Mario, and Peter Galison, eds. *Scientific Authorship: Credit and Intellectual Property in Science*. New York: Routledge, 2002.

Bianco, Paolo, and Douglas Sipp. "Regulation: Sell Help Not Hope." *Nature* 510, no. 7505 (June 19, 2014): 336.

Binkley, Sam. *Happiness as Enterprise: An Essay on Neoliberal Life*. Albany: SUNY Press, 2014.

Bishop, Jeffrey. *The Anticipatory Corpse: Medicine, Power, and the Care of the Dying*. Notre Dame: University of Notre Dame Press, 2011.

Blumenberg, Hans. *The Legitimacy of the Modern Age*. Cambridge, MA: MIT Press, 1985.

Blundell, Mary Whitlock. "Parental Nature and Stoic Oikeiosis." *Ancient Philosophy* 10 (1990): 221–42.

Boltanski, Luc, and Eve Chiapello. *The New Spirit of Capitalism*, trans. Gregory Elliott. New York: Verso, 2005.

Bourdieu, Pierre. *The Logic of Practice*, trans. Richard Nice. Stanford, CA: Stanford University Press, 1992.

Bourdieu, Pierre, and Loïc J. D. Wacquant. *An Invitation to Reflexive Sociology*. Chicago: University of Chicago Press, 1992.

Bowlin, John. *Contingency and Fortune in Aquinas's Ethics*. New York: Cambridge University Press, 1999.

Breen, Keith. "The State, Compartmentalization and the Turn to Local Community: A Critique of the Political Thought of Alasdair MacIntyre." *The European Legacy* 10, no. 5 (August 2005): 485–501.

Under Weber's Shadow: Modernity, Subjectivity and Politics in Habermas, Arendt and Macintyre. Burlington, VT: Ashgate, 2012.

Brock, Brian. *Christian Ethics in a Technological Age*. Grand Rapids, MI: William Eerdmans, 2010.

Brooke, John Hedley. *Science and Religion: Some Historical Perspectives*. New York: Cambridge University Press, 1991.

Brown, Peter. *Power and Persuasion in Late Antiquity: Towards a Christian Empire*. Madison: University of Wisconsin Press, 1992.

Brown, Wendy. *Undoing the Demos: Neoliberalism's Stealth Revolution*. New York: Zone Books, 2015.

Burchell, David. "Civic Personae: MacIntyre, Cicero and Moral Personality." *History of Political Thought* 19, no. 1 (1998): 101–18.

Burns, Tony. "Revolutionary Aristotelianism? The Political Thought of Aristotle, Marx, and MacIntyre." In *Virtue and Politics: Alasdair MacIntyre's Revolutionary*

Aristotelianism, ed. Paul Blackledge and Kelvin Knight, 35–53. Notre Dame, IN: University of Notre Dame Press, 2011.

Byers, Sarah Catherine. *Perception, Sensibility, and Moral Motivation in Augustine: A Stoic-Platonic Synthesis*. New York: Cambridge University Press, 2013.

Campos, Francisco, et al. "Teaching Personal Initiative Beats Traditional Training in Boosting Small Business in West Africa." *Science* 357, no. 6357 (September 22, 2017): 1287–90.

Carreyrou, John. "Hot Startup Theranos Has Struggled with Its Blood-Test Technology." *Wall Street Journal*, October 16, 2015.

Carreyrou, John, Michael Siconolfi, and Christopher Weaver. "Theranos Dealt Sharp Blow as Elizabeth Holmes Is Banned from Operating Labs." *Wall Street Journal*, July 8, 2016, sec. Business.

Casadevall, Arturo, and Ferric C. Fang. "Reforming Science: Methodological and Cultural Reforms." *Infection and Immunity* 80, no. 3 (2012): 891–6.

"Winner Takes All." *Scientific American* 307, no. 2 (2012): 13.

Cattaneo, Elena, and Gilberto Corbellini. "Stem Cells: Taking a Stand against Pseudoscience." *Nature* 510, no. 7505 (June 16, 2014) 333–5.

Cavadini, John. "The Darkest Enigma: Reconsidering the Self in Augustine's Thought." *Augustinian Studies* 38, no. 1 (2007): 119–32.

"The Structure and Intention of Augustine's De Trinitate." *Augustinian Studies* 23 (1992): 103–23.

Chargaff, Erwin. *Heraclitean Fire: Sketches from a Life before Nature*. New York: Rockefeller University Press, 1978.

Cicero, Marcus Tullius. *On Moral Ends*. New York: Cambridge University Press, 2001.

Tusculan Disputations, trans. J. E. King. Loeb Classical Library 141. Cambridge, MA: Harvard University Press, 1927.

Cloutier, David, and William Mattison III. "The Resurgence of Virtue in Recent Moral Theology." *Journal of Moral Theology* 3, no. 1 (2014): 228–59.

Cochran, Elizabeth Agnew. "Bricolage and the Purity of Traditions: Engaging the Stoics for Contemporary Christian Ethics." *Journal of Religious Ethics* 40, no. 4 (2012): 720–9.

"Consent, Conversion, and Moral Formation: Stoic Elements in Jonathan Edwards's Ethics." *Journal of Religious Ethics* 39, no. 4 (2011): 623–50.

Protestant Virtue and Stoic Ethics. London: Bloomsbury, 2017.

Colish, Marcia L. *The Stoic Tradition from Antiquity to the Early Middle Ages*, 2 vols. New York: E. J. Brill, 1990.

Collins, Francis S. *The Language of God: A Scientist Presents Evidence for Belief*. New York: Free Press, 2007.

Collins, Harry. *Changing Order: Replication and Induction in Scientific Practice*. London: Sage Publications, 1985.

Collins, Harry, and Robert Evans. "The Third Wave of Science Studies." *Social Studies of Science* 32, no. 2 (2002).

Congregation for the Doctrine of the Faith. *Instructions Dignitas Personae on Certain Bioethical Questions*. Nairobi: Pauline Publications Africa, 2009.

"Contaminated Cell Lines Throw Thousands of Studies into Question." Stat. July 21, 2016, www.statnews.com/2016/07/21/studies-wrong-cells/.

Cooper, John. "Eudaimonism and the Appeal to Nature in the Morality of Happiness: Comments on Julia Annas, The Morality of Happiness." *Philosophy and Phenomenological Research* 55, no. 3 (1997): 587–98.

"Socrates and Philosophy as a Way of Life." In *Maieusis: Essays in Honor of Myles Burnyeat*, ed. Dominic Scott. New York: Oxford University Press, 2007.

Cooper, Melinda. *Life as Surplus: Biotechnology and Capitalism in the Neoliberal Era*. Seattle: University of Washington Press, 2008.

Cooper, Melinda, and Catherine Waldby. *Clinical Labor: Tissue Donors and Research Subjects in the Global Bioeconomy*. Durham, NC: Duke University Press, 2014.

Coquin, R. G. "La Theme de la Parrhesia et Ses Expressions Symboliques dans les Rites D'initiation a Antioche." *Proche Orient Chretien* 20 (1970): 3–19.

Cover, Michael. *Lifting the Veil*. Berlin: De Gruyter, 2015.

Coyne, Jerry. "Francis Collins as NIH Director." Why Evolution Is True. Accessed August 20, 2013. https://whyevolutionistrue.wordpress.com/2009/07/10/francis-collins-as-nih-director/.

Cyranoski, David. "Stem Cells in Texas: Cowboy Culture." *Nature* 494, no. 7436 (February 14, 2013): 166.

Dalton, Rex, and Quirin Schiermeier. "Genentech Pays $200m over Growth Hormone 'Theft.'" *Nature* 402 (1999): 335.

Daniels, Ronald. "A Generation at Risk: Young Investigators and the Future of the Biomedical Workforce." *Proceedings of the National Academy of Sciences* 112, no. 2 (2015): 313–18.

Davidson, Arnold I. "Ethics as Ascetics: Foucault, the History of Ethics, and Ancient Thought." In *The Cambridge Companion to Foucault*, ed. Gary Gutting. New York: Cambridge University Press, 1994.

de Certeau, Michel. *The Practice of Everyday Life*. Berkeley: University of California Press, 2011.

de Chadarevian, Soraya. "Interview with Sydney Brenner." *Studies in History and Philosophy of Science Part C: Studies in History and Philosophy of Biological and Biomedical Sciences* 40, no. 1 (March 2009): 65–71.

Dean, Mitchell. *Governmentality: Power and Rule in Modern Society*. Thousand Oaks, CA: Sage Publications, 1999.

Deane-Drummond, Celia. *Christ and Evolution: Wonder and Wisdom*. Minneapolis, MN: Fortress Press, 2009.

Genetics and Christian Ethics. New York: Cambridge University Press, 2005.

del Nido, Daniel. "Pierre Hadot on Habit, Reason, and Spiritual Exercises." *Journal of Religious Ethics* 46, no. 1 (March 1, 2018): 7–36.

Deltete, Robert J. "Man of Science, Man of Faith: Pierre Duhem's 'Physique de Croyant.'" *Zygon* 43, no. 3 (2008): 627–37.

Dougherty, Conor. "They Promised Us Jet Packs. They Promised the Bosses Profit." *New York Times*, July 23, 2016, www.nytimes.com/2016/07/24/technology/they-promised-us-jet-packs-they-promised-the-bosses-profit.html.

Drahos, Peter, and John Braithwaite. *Information Feudalism: Who Owns the Knowledge Economy?* New York: New Press, 2003.

Dugdale, Lydia, ed., *Dying in the Twenty-First Century: Toward a New Ethical Framework for the Art of Dying Well.* Cambridge, MA: MIT Press, 2016.

Duhem, Pierre. *The Aim and Structure of Physical Theory*. Princeton, NJ: Princeton University Press, 1954.

"Physics of a Believer." In *The Aim and Structure of Physical Theory*, 273–311. Princeton, NJ: Princeton University Press, 1954.

To Save the Phenomena: An Essay on the Idea of Physical Theory from Plato to Galileo. Chicago: University of Chicago Press, 1969.

Dumit, Joseph. *Drugs for Life: How Pharmaceutical Companies Define Our Health.* Durham, NC: Duke University Press, 2012.

Editors of Lingua Franca. *The Sokal Hoax: The Sham That Shook the Academy.* Lincoln: University of Nebraska Press, 2000.

Edmond, Gary. "Supersizing Daubert: Science for Litigation and Its Implications for Legal Practice and Scientific Research." *Villanova Law Review* 52 (2007): 857.

Ehrenreich, Barbara. *Bright-Sided: How Positive Thinking Is Undermining America*. New York: Picador, 2010.

Elliott, Carl. *White Coat, Black Hat: Adventures on the Dark Side of Medicine.* Boston: Beacon Press, 2011, 25–50.

Elliot, David. *Hope and Christian Ethics*. New York: Cambridge University Press, 2017.

Ellul, Jacques. *The Technological Society*, trans. John Wilkinson. New York: Vintage Books, 1964.

Emery, Gilles. *Trinity, Church, and the Human Person: Thomistic Essays*. Naples, FL: Sapientia Press, 2007.

Engberg-Pedersen, Troels. *Cosmology and Self in the Apostle Paul: The Material Spirit*. Oxford: Oxford University Press, 2010.

Paul and the Stoics. Louisville, KY: Westminster John Knox Press, 2000.

Epictetus. *Discourses, Books 1–2*, trans. W. A. Oldfather. Loeb Classical Library 131. Cambridge, MA: Harvard University Press, 1998.

Fanelli, Daniele. "How Many Scientists Fabricate and Falsify Research? A Systematic Review and Meta-Analysis of Survey Data." *PLOS ONE* 4, no. 5 (May 29, 2009): e5738.

Fang, Ferric C., and Arturo Casadevall. "Reforming Science: Structural Reforms." *Infection and Immunity* 80, no. 3 (March 1, 2012): 897–901.

Fang, Ferric C., R. Grant Steen, and Arturo Casadevall. "Misconduct Accounts for the Majority of Retracted Scientific Publications." *Proceedings of the National Academy of Sciences* 109, no. 42 (October 1, 2012): 17028–33.

Fergusson, David. *Community, Liberalism and Christian Ethics*. New York: Cambridge University Press, 1999.

Finnis, John. *Natural Law and Natural Rights*, 2nd edition. New York: Oxford University Press, 2011.

Fisher, Jill. *Medical Research for Hire: The Political Economy of Pharmaceutical Clinical Trials*. New Brunswick, NJ: Rutgers University Press, 2008.

Fitzgerald, John T., ed. *Friendship, Flattery, and Frankness of Speech: Studies on Friendship in the New Testament World*. New York: Brill, 1996.

Fleck, Ludwik. *Genesis and Development of a Scientific Fact*. Chicago: University of Chicago Press, 1979.

Flett, John. "Alasdair MacIntyre's Tradition-Constituted Enquiry in Polanyian Perspective." *Tradition and Discovery: The Polanyi Society Periodical* 26, no. 2 (2009): 6–20.

Foot, Philippa. *Natural Goodness*. Oxford: Clarendon Press, 2003.

Foucault, Michel. *The Archaeology of Knowledge*. New York: Pantheon Books, 1972.

The Birth of Biopolitics: Lectures at the Collège de France, 1978–79, ed. Michel Senellart. New York: Palgrave Macmillan, 2008.

The Courage of Truth: The Government of Self and Others II: Lectures at the Collège de France 1983–1984, trans. Graham Burchell. New York: Palgrave Macmillan, 2011.

Discipline and Punish: The Birth of the Prison. New York: Pantheon Books, 1977.

Du Gouvernement des Vivants: Cours au Collège de France (1979–1980), ed. François Ewald, Alessandro Fontana, and Michel Senellart. Paris: Seuil, 2012.

"The Ethics of the Concern for Self as a Practice of Freedom." In *Ethics, Subjectivity, and Truth*, ed. Paul Rabinow, 281–302. New York: New Press, 1998.

"On the Genealogy of Ethics: An Overview of Work in Progress." In *Ethics, Subjectivity, and Truth*, ed. Paul Rabinow, 253–80. New York: New Press, 1998.

The Government of Self and Others. New York: Palgrave Macmillan, 2010.

The Hermeneutics of the Subject: Lectures at the Collège de France, 1981–82, trans. Graham Burchell. New York: Palgrave Macmillan, 2005.

The History of Sexuality, vol. 1. New York: Pantheon Books, 1978.

"Introduction." In *The Normal and the Pathological*, by Georges Canguilhem, 7–24. New York: Zone Books, 1989.

Psychiatric Power: Lectures at the College de France 1973–1974, ed. Arnold I. I. Davidson and Graham Burchell. Basingstoke: Palgrave Macmillan, 2006.

Security, Territory, Population: Lectures at the Collège de France, 1977–78, ed. Michel Senellart, trans. Graham Burchell. New York: Palgrave Macmillan, 2007.

Society Must Be Defended: Lectures at the Collège de France, 1975–76, ed. Mauro Bertani and Alessandro Fontana, trans. David Macey. New York: Picador, 2003.

"The Subject and Power." In *Power*, ed. James Faubion, 326–48. New York: New Press, 2001.

"Truth and Power." In *Power*, ed. James Faubion, 111–33. New York: New Press, 2001.

The Use of Pleasure. New York: Vintage Books, 1985.
"What Is Enlightenment?" In *Ethics, Subjectivity, and Truth*, ed. Paul Rabinow, 303–20. New York: New Press, 1998.

Foucault, Michel, Graham Burchell, Colin Gordon, and Peter Miller. *The Foucault Effect: Studies in Governmentality*. Chicago: University of Chicago Press, 1991.

Francis. *Laudato si'*. 2015. http://w2.vatican.va/content/francesco/en/encyclicals/documents/papa-francesco_20150524_enciclica-laudato-si.html.

Fraser, Nancy. "Foucault on Modern Power: Empirical Insights and Normative Confusions." *Praxis International*, no. 3 (1981): 272–87.

Freedman, Leonard P., Iain M. Cockburn, and Timothy S. Simcoe. "The Economics of Reproducibility in Preclinical Research." *PLOS Biology* 13, no. 6 (June 9, 2015): e1002165.

Fuller, Steve. *Thomas Kuhn: A Philosophical History for Our Times*. Chicago: University of Chicago Press, 2000.

Gertner, Jon. *The Idea Factory: Bell Labs and the Great Age of American Innovation*. New York: Penguin Books, 2013.

Gill, Christopher. *The Structured Self in Hellenistic and Roman Thought*. New York: Oxford University Press, 2006.

Gioia, Luigi. *The Theological Epistemology of Augustine's De Trinitate*. Oxford: Oxford University Press, 2009.

Golan, Tal. "Revisiting the History of Scientific Expert Testimony." *Brooklyn Law Review* 73 (2008): 933.

Gold, Michael. *A Conspiracy of Cells*. Albany: State University of New York Press, 1985.

Goodstein, David. *On Fact and Fraud: Cautionary Tales from the Front Lines of Science*. Princeton, NJ: Princeton University Press, 2010.

Goodyear, Dana. "The Stem-Cell Scandal," *The New Yorker*, February 22, 2016, www.newyorker.com/magazine/2016/02/29/the-stem-cell-scandal.

Grant, George. "Knowing and Making." In *The George Grant Reader*, ed. William Christian and Sheila Grant, 407–17. Toronto: University of Toronto Press, 1998.

Graver, Margaret. *Stoicism and Emotion*. Chicago: University of Chicago Press, 2007.

Greene, Jeremy A. *Prescribing by Numbers: Drugs and the Definition of Disease*. Baltimore, MD: Johns Hopkins University Press, 2008.

Gregory Nazianen. "On St. Basil the Great, Bishop of Caesarea." In *Funeral Orations*, trans. Leo McCauley, 27–100. The Fathers of the Church, v. 22. New York: Fathers of the Church, 1953.

Grene, Marjorie. "Response to Alasdair MacIntyre." In *Morals, Science, and Sociality*, ed. H. Tristram Englehardt and Daniel Callahan, 40–47. Hastings-on-Hudson, NY: Hastings Center, 1978.

Grewal, Seema. "An Interview with Didier Stainier." *Development* 142, no. 17 (September 1, 2015): 2861–3.

Gutting, Gary. *Michel Foucault's Archaeology of Scientific Reason: Science and the History of Reason*. New York: Cambridge University Press, 1989.

Hacking, Ian. *The Social Construction of What?* Cambridge, MA: Harvard University Press, 2000.

Hadot, Ilsetraut. *Sénèque: Direction spirituelle et pratique de la philosophie*. Paris: Vrin, 2014.

Hadot, Pierre. *Philosophy as a Way of Life: Spiritual Exercises from Socrates to Foucault*, trans. Arnold I. Davidson. Malden, MA: Blackwell, 1995.

Halperin, David M. *Saint Foucault: Towards a Gay Hagiography*. New York: Oxford University Press, 1995.

Harris, Richard. *Rigor Mortis: How Sloppy Science Creates Worthless Cures, Crushes Hope, and Wastes Billions*. New York: Basic Books, 2017.

Harrison, Peter. *The Territories of Science and Religion*. Chicago: University of Chicago Press, 2017.

Harvey, David. *A Brief History of Neoliberalism*. Oxford: Oxford University Press, 2007.

The Condition of Postmodernity: An Enquiry into the Origins of Cultural Change. Cambridge, MA: Blackwell, 1990.

Harvey, John. "The Nature of the Infused Moral Virtues." *Proceedings of the Tenth Annual Convention of the Catholic Theological Society of America* 10 (1955): 172–221.

Hauerwas, Stanley. *A Community of Character: Toward a Constructive Christian Social Ethic*. Notre Dame, IN: University of Notre Dame Press, 1991.

Hayek, Friedrich A. von. *Individualism and Economic Order*. Chicago: University of Chicago Press, 1948.

New Studies in Philosophy, Politics, Economics and the History of Ideas. Chicago: University of Chicago Press, 1978.

Heidegger, Martin. "The Age of the World Picture." In *The Question Concerning Technology, and Other Essays*, 115–54. New York: Harper Torchbooks, 1977.

"The Question Concerning Technology." In *The Question Concerning Technology, and Other Essays*, 3–35. New York: Harper Torchbooks, 1977.

Herdt, Jennifer A. *Putting on Virtue: The Legacy of the Splendid Vices*. Chicago: University of Chicago Press, 2012.

"The Virtue of the Liturgy." In *The Blackwell Companion to Christian Ethics*, ed. Stanley Hauerwas and Samuel Wells. Oxford: Wiley-Blackwell, 2011.

Hibbs, Thomas. "MacIntyre's Postmodern Thomism: Reflections on Three Rival Versions of Moral Enquiry." *The Thomist* 57, no. 2 (1993): 277.

Hicks, Diana, Paul Wouters, Ludo Waltman, et al. "Bibliometrics: The Leiden Manifesto for Research Metrics." *Nature* 520, no. 7548 (April 23, 2015): 429.

Hierocles. *Hierocles the Stoic: Elements of Ethics, Fragments and Excerpts*, trans. Ilaria Ramelli. Atlanta, GA: Society of Biblical Literature, 2009.

Hiltzik, Michael. "California Moves against Unlicensed Stem Cell Treatments – But Is It Doing Enough?" *Los Angeles Times*, January 25, 2018. www.latimes.com/business/hiltzik/la-fi-hiltzik-stemcell-regulation-20171013-story.html.

Hitchens, Christopher. *God Is Not Great: How Religion Poisons Everything.* New York: Twelve, 2009.

Hovey, Craig. *Bearing True Witness: Truthfulness in Christian Practice.* Grand Rapids, MI: Eerdmans, 2011.

Huber, Peter. *Galileo's Revenge.* New York: Basic Books, 1993.

Hume, David. *A Treatise of Human Nature: A Critical Edition*, ed. David Fate Norton and Mary J. Norton. New York: Clarendon Press, 2011.

Husserl, Edmund. *The Crisis of European Sciences and Transcendental Phenomenology.* Evanston, IL: Northwestern University Press, 1970.

International Theological Commission. "Communion and Stewardship: Human Persons Created in the Image of God," 2004. www.vatican.va/roman_curia/congregations/cfaith/cti_documents/rc_con_cfaith_doc_20040723_communion-stewardship_en.html.

Inwood, Brad. *Reading Seneca: Stoic Philosophy at Rome.* Oxford: Clarendon Press, 2005.

Ioannidis, John P. A. "Why Most Published Research Findings Are False." *PLOS Medicine* 2, no. 8 (August 30, 2005): e124.

Ioannidis, John P. A., Michael Stuart, Shannon Brownlee, and Sheri Strite. "How to Survive the Medical Misinformation Mess." *European Journal of Clinical Investigation* 47, no. 11 (2017): 795–802.

Jasanoff, Sheila. "Breaking the Waves in Science Studies." *Social Studies of Science* 33, no. 3 (2003): 389–400.

"Transparency in Public Science: Purposes, Reasons, Limits." *Law and Contemporary Problems* 68 (2006): 21–45.

John Paul II. "Centesimus Annus." In *Catholic Social Thought: The Documentary Heritage*, updated/expanded edition, ed. David J. O'Brien and Thomas A. Shannon, 471–523. Maryknoll, NY: Orbis Books, 2010.

Johnson, Carolyn Y. "Stem Cell Scientist Sues Harvard for Misconduct Investigation." *Boston Globe*, December 17, 2014.

Josephson-Storm, Jason A. *The Myth of Disenchantment: Magic, Modernity, and the Birth of the Human Sciences.* Chicago: University of Chicago Press, 2017.

Kaiser, Jocelyn. "Janelia Farm to Recruit First Class." *Science* 306, no. 5694 (2004): 210–11.

Kant, Immanuel. *Religion within the Bounds of Bare Reason.* Indianapolis, IN: Hackett, 2009.

Kaufman, Sharon R. *Ordinary Medicine: Extraordinary Treatments, Longer Lives, and Where to Draw the Line.* Durham, NC: Duke University Press, 2015.

Kay, Lily. *Who Wrote the Book of Life?: A History of the Genetic Code.* Stanford, CA: Stanford University Press, 2000.

Kevles, Daniel. *In the Name of Eugenics: Genetics and the Uses of Human Heredity.* Cambridge, MA: Harvard University Press, 1995.

Knight, Kelvin. *Aristotelian Philosophy: Ethics and Politics from Aristotle to MacIntyre.* Cambridge: Polity, 2007.

"Revolutionary Aristotelianism." In *Virtue and Politics: Alasdair MacIntyre's Revolutionary Aristotelianism*, ed. Paul Blackledge and Kelvin Knight, 20–34. Notre Dame, IN: University of Notre Dame Press, 2011.

Knobel, A. "Can Aquinas's Infused and Acquired Virtues Coexist in the Christian Life?" *Studies in Christian Ethics* 23, no. 4 (2010): 381–96.

Kolbet, Paul R. *Augustine and the Cure of Souls: Revising a Classical Ideal.* Notre Dame, IN: University of Notre Dame Press, 2009.

Kooten, Geurt Hendrik van. *Paul's Anthropology in Context: The Image of God, Assimilation to God, and Tripartite Man in Ancient Judaism, Ancient Philosophy and Early Christianity.* Tübingen: Mohr Siebeck, 2008.

Krimsky, Sheldon. *Science in the Private Interest: Has the Lure of Profits Corrupted Biomedical Research?* Lanham, MD: Rowman and Littlefield, 2003.

Krimsky, Sheldon, James G. Ennis, and Robert Weissman. "Academic-Corporate Ties in Biotechnology: A Quantitative Study." *Science, Technology, & Human Values* 16, no. 3 (July 1, 1991): 275–87. https://doi.org/10.2307/689916.

Kruger, Matthew. "Aquinas, Hadot, and Spiritual Exercises." *New Blackfriars* 98, no. 1076 (July 1, 2017): 414–26.

Kuhn, Thomas. *The Structure of Scientific Revolutions*, 3rd edition. Chicago: University of Chicago Press, 1996.

LaCugna, Catherine. *God for Us: The Trinity and Christian Life.* San Francisco, CA: Harper, 1993.

Laidlaw, James. *The Subject of Virtue: An Anthropology of Ethics and Freedom.* New York: Cambridge University Press, 2014.

Landecker, Hannah. *Culturing Life: How Cells Became Technologies.* Cambridge, MA: Harvard University Press, 2007.

Landsberg, Mitchell. "Vatican Signs Deal to Collaborate on Adult Stem Cell Research." *Los Angeles Times*, October 20, 2011. http://articles.latimes.com/2011/oct/20/business/la-fi-vatican-stem-cells-20111020.

Latour, Bruno. *Science in Action: How to Follow Scientists and Engineers through Society.* Cambridge, MA: Harvard University Press, 1988.

"Why Has Critique Run Out of Steam? From Matters of Fact to Matters of Concern." *Critical Inquiry* 39 (2004): 225–48.

Lear, Gabriel. *Happy Lives and the Highest Good: An Essay on Aristotle's Nicomachean Ethics.* Princeton, NJ: Princeton University Press, 2004.

Ledford, Heidi. "Bitter CRISPR Patent War Intensifies." *Nature News*, 10.1038/nature.2017.22892.

Leon, Sharon. *An Image of God: The Catholic Struggle with Eugenics.* Chicago: University of Chicago Press, 2013.

Lewis, Sinclair. *Arrowsmith.* New York: Harcourt Brace Jovanovich, 1952.

Lewontin, Richard, and Richard Levins. "Stephen Jay Gould – What Does It Mean to Be a Radical?" In *Stephen Jay Gould: Reflections on His View of Life*, ed. Patricia Kelley and Robert Ross, 199–205. New York: Oxford University Press, 2009.

Long, A. A. "Greek Ethics after MacIntyre and the Stoic Community of Reason." *Ancient Philosophy* 3 (1983): 184–97.

Hellenistic Philosophy; Stoics, Epicureans, Sceptics. New York: Scribner, 1974.

Long, A. A., and D. N Sedley. *The Hellenistic Philosophers.* New York: Cambridge University Press, 1987.

Lundh, Andreas, Joel Lexchin, Barbara Mintzes, et al. "Industry Sponsorship and Research Outcome." *The Cochrane Database of Systematic Reviews* 12 (December 12, 2012): MR000033.

MacIntyre, Alasdair. *After Virtue: A Study in Moral Theory*, 2nd edition. Notre Dame, IN: University of Notre Dame Press, 1984.

Alasdair MacIntyre's Engagement with Marxism: Selected Writings 1953–1974. Boston: Brill, 2008.

Dependent Rational Animals: Why Human Beings Need the Virtues. Chicago: Open Court, 1999.

Edith Stein: A Philosophical Prologue, 1913–1922. Lanham, MD: Rowman & Littlefield, 2006.

"The End of Education." *Commonweal*, October 20, 2006.

"Epistemological Crises and Dramatic Narrative." In *The Tasks of Philosophy: Selected Essays*, vol. 1: 3–23. Cambridge: Cambridge University Press, 2006.

Ethics in the Conflicts of Modernity: An Essay on Desire, Practical Reasoning, and Narrative. New York: Cambridge University Press, 2016.

"First Principles, Final Ends, and Contemporary Philosophical Issues." In *The Tasks of Philosophy: Selected Essays*, vol. 1: 143–78. New York: Cambridge University Press, 2006.

Herbert Marcuse: An Exposition and a Polemic. New York: Viking Press, 1970.

"An Interview with Cogito." In *The MacIntyre Reader*, ed. Kelvin Knight, 267–75. Notre Dame, IN: University of Notre Dame Press, 1998.

"Marxist Mask and Romantic Face: Lukács on Thomas Mann." In *Alasdair MacIntyre's Engagement with Marxism: Selected Writings 1953–1974*, ed. Paul Blackledge and Neil Davidson, 317–27. Boston: Brill, 2008.

"Objectivity in Morality and Objectivity in Science." In *Morals, Science, and Sociality*, ed. H. Tristram Englehardt and Daniel Callahan, 21–39. Hastings-on-Hudson, NY: Hastings Center, 1978.

A Short History of Ethics: A History of Moral Philosophy from the Homeric Age to the Twentieth Century, 2nd edition. Notre Dame, IN: University of Notre Dame Press, 1998.

"Social Structures and Their Threats to Moral Agency." *Philosophy* 74, no. 3 (1999): 311–29.

"The Theses on Feuerbach: A Road Not Taken." In *The MacIntyre Reader*, ed. Kelvin Knight, 223–34. Notre Dame, IN: University of Notre Dame Press, 1998.

Three Rival Versions of Moral Enquiry: Encyclopaedia, Genealogy, and Tradition. Notre Dame, IN: University of Notre Dame Press, 1990.

Whose Justice? Which Rationality? Notre Dame, IN: University of Notre Dame Press, 1988.

Mackey, James. *Power and Christian Ethics*. New York: Cambridge University Press, 1994.

Mahmood, Saba. *Politics of Piety: The Islamic Revival and the Feminist Subject*, 2nd revised edition. Princeton, NJ: Princeton University Press, 2005.

Mahoney, John. *The Making of Moral Theology: A Study of the Roman Catholic Tradition*. New York: Oxford University Press, 1989.

March for Science. "The Science behind the March for Science Crowd Estimates." *Medium* (blog), May 15, 2017, https://medium.com/marchforscience-blog/the-science-behind-the-march-for-science-crowd-estimates-f337adf2d665.

Marcus Aurelius. *Meditations*, trans. Robin Hard. New York: Oxford University Press, 2011.

Marcuse, Herbert. *One-Dimensional Man: Studies in the Ideology of Advanced Industrial Society*, 2nd revised edition. London: Routledge, 2002.

Marrow, Stanley. "Parrhesia and the New Testament." *The Catholic Biblical Quarterly* 44 (1982): 431–46.

Martin, Emily. *Flexible Bodies*. Boston: Beacon Press, 1995.

Martin, R. N. D. *Pierre Duhem: Philosophy and History in the Work of a Believing Physicist*. La Salle, IL: Open Court, 1991.

Marx, Karl. "Theses on Feuerbach." In *The Marx-Engels Reader*, ed. Robert Tucker, 107–9. New York: Norton, 1972.

Matthews, Dave. "Decrease in UK Grant Success Rates Prompts Worldwide Comparisons." *Times Higher Education Supplement*, October 6, 2016.

Mattison, William III. "Can Christians Possess the Acquired Cardinal Virtues?" *Theological Studies* 72, no. 3 (2011): 558–85.

May, Christopher, and Susan K. Sell. *Intellectual Property Rights: A Critical History*. Boulder, CO: Lynne Rienner, 2006.

Mayr, Ernst. "The Multiple Meanings of Teleological." In *Toward a New Philosophy of Biology: Observations of an Evolutionist*, 38–66. Cambridge, MA: Harvard University Press, 1989.

McGarity, Thomas O., and Wendy E. Wagner. *Bending Science: How Special Interests Corrupt Public Health Research*. Cambridge, MA: Harvard University Press, 2008.

McGrath, Alister E. *Dawkins' God: From the Selfish Gene to the God Delusion*. Malden, MA: Wiley, 2015.

McKenny, Gerald. *Biotechnology, Human Nature, and Christian Ethics*. New York: Cambridge University Press, 2018.

To Relieve the Human Condition: Bioethics, Technology, and the Body. Albany: State University of New York Press, 1997.

Merton, Robert K. *The Sociology of Science*. Chicago: University of Chicago Press, 1973.

Mervis, Jeffrey. "Business Decisions." *Science* 328, no. 6240 (2015): 1190–3.

Miller, Jim. *The Passion of Michel Foucault*. New York: Simon & Schuster, 1993.

Mirowski, Philip. "Defining Neoliberalism." In *The Road from Mont Pelerin: The Making of the Neoliberal Thought Collective*, ed. Philip Mirowski and Dieter Plehwe, 417–56. Cambridge, MA: Harvard University Press, 2009.

Never Let a Serious Crisis Go to Waste. New York: Verso, 2013.

Science-Mart: Privatizing American Science. Cambridge, MA: Harvard University Press, 2011.

Mitchell, Mark T. "Michael Polanyi, Alasdair MacIntyre, and the Role of Tradition." *Humanitas* 19, no. 1 (2006): 97–125.

Moore, G. E. *Principia Ethica*. Cambridge: Cambridge University Press, 1922.

Moss, Lenny. *What Genes Can't Do*. Cambridge, MA: MIT Press, 2004.

Nagel, Thomas. *Mind and Cosmos: Why the Materialist Neo-Darwinian Conception of Nature Is Almost Certainly False*. New York: Oxford University Press, 2012.

"National Bioeconomy Blueprint." The White House, April 2012. www.whitehouse.gov/sites/default/files/microsites/ostp/national_bioeconomy_blueprint_april_2012.pdf.

National Academies of Sciences, Engineering, and Medicine. *Fostering Integrity in Research*. Washington, DC: National Academies Press, 2017.

Statistical Challenges in Assessing and Fostering the Reproducibility of Scientific Results: Summary of a Workshop. Washington, DC: National Academies Press, 2016.

National Institutes of Health. *Biomedical Research Workforce Working Group Report*. Bethesda, MD: National Institutes of Health, 2012.

National Research Council. *Research Training in the Biomedical, Behavioral, and Clinical Research Sciences*. Washington, DC: National Academies Press, 2011.

Needleman, Herbert. "Salem Comes to the National Institutes of Health: Notes from Inside the Crucible of Scientific Integrity." *Pediatrics* 90, no. 6 (1992): 977–81.

Nehamas, Alexander. *The Art of Living: Socratic Reflections from Plato to Foucault*. Berkeley: University of California Press, 1998.

Newman, John Henry. *An Essay on Development of Christian Doctrine*. Notre Dame, IN: University of Notre Dame Press, 1989.

Nightingale, Paul, and Paul Martin. "The Myth of the Biotech Revolution." *Trends in Biotechnology* 22, no. 11 (November 2004): 564–9.

Normile, Dennis, Gretchen Vogel, and Constance Holden. "Cloning Researcher Says Work Is Flawed but Claims Results Stand." *Science* 310, no. 5756 (2005): 1886–7.

Northcott, Michael S. *The Environment and Christian Ethics*. New York: Cambridge University Press, 1996.

Nowak, Martin A., and Sarah Coakley, eds. *Evolution, Games, and God: The Principle of Cooperation*. Cambridge, MA: Harvard University Press, 2013.

Nussbaum, Martha. *The Therapy of Desire: Theory and Practice in Hellenistic Ethics*. Princeton, NJ: Princeton University Press, 1996.

Nye, Mary Jo. *Michael Polanyi and His Generation: Origins of the Social Construction of Science*. Chicago: University of Chicago Press, 2011.

Obokata, Haruko, Yoshiki Sasai, Hitoshi Niwa, et al. "Bidirectional Developmental Potential in Reprogrammed Cells with Acquired Pluripotency." *Nature* 505, no. 7485 (January 29, 2014): 676.

Obokata, Haruko, Teruhiko Wakayama, Yoshiki Sasai, et al. "Stimulus-Triggered Fate Conversion of Somatic Cells into Pluripotency." *Nature* 505, no. 7485 (January 29, 2014): 641.

O'Donovan, Oliver. *The Problem of Self-Love in St. Augustine*. Eugene, OR: Wipf and Stock, 2006.

Office of the Commissioner. "Press Announcements – FDA Warns US Stem Cell Clinic of Significant Deviations." January, 25, 2018. www.fda.gov/NewsEvents/Newsroom/PressAnnouncements/ucm573431.htm.

Open Science Collaboration. "Estimating the Reproducibility of Psychological Science." *Science* 349, no. 6251 (August 28, 2015): aac4716.

Oreskes, Naomi, and Erik Conway. *Merchants of Doubt: How a Handful of Scientists Obscured the Truth on Issues from Tobacco Smoke to Global Warming.* New York: Bloomsbury Press, 2010.

Paras, Eric. *Foucault 2.0: Beyond Power and Knowledge.* New York: Other Press, 2006.

Pauly, Philip. *Controlling Life: Jacques Loeb and the Engineering Ideal in Biology.* New York: Oxford University Press, 1987.

Payne, Christopher. *The Consumer, Credit and Neoliberalism: Governing the Modern Economy.* New York: Routledge, 2012.

Peng, Roger. "The Reproducibility Crisis in Science: A Statistical Counterattack." *Significance* 12, no. 3 (June 1, 2015): 30–2.

Peters, Ted. "Stem Cells, Aging, and the Fountain of Youth." *Theology and Science* 10, no. 3 (2012): 221–3.

Pinckaers, Servais. *The Sources of Christian Ethics*, 3rd edition, trans. Mary Thomas Noble. Washington, DC: Catholic University of America Press, 1995.

Pinkard, Terry. "MacIntyre's Critique of Modernity." In *Alasdair MacIntyre*, ed. Mark Murphy, 176–200. New York: Cambridge University Press, 2003.

Plato. "Alcibiades I." In *Complete Works*, ed. John Cooper, trans. D. S. Hutchinson, 557–95. Indianapolis, IN: Hackett, 1997.

"Laches." In *Complete Works*, ed. John Cooper, trans. Rosamund Sprague. Indianapolis, IN: Hackett, 1997.

Polanyi, Michael. "Patent Reform." *Review of Economic Studies* 11, no. 2 (1944): 61–76.

Personal Knowledge: Towards a Post-Critical Philosophy. Chicago: University of Chicago Press, 1962.

"The Republic of Science: Its Political and Economic Theory." *Minerva* 38, no. 1 (2000): 1–21.

The Tacit Dimension. Reissue. Chicago: University of Chicago Press, 2009.

Pollack, Robert. *The Faith of Biology and the Biology of Faith.* New York: Columbia University Press, 2000.

Pope, Stephen J. *Human Evolution and Christian Ethics.* New York: Cambridge University Press, 2007.

Porter, Jean. "Tradition in the Recent Work of Alasdair MacIntyre." In *Alasdair MacIntyre*, ed. Mark Murphy, 38–69. New York: Cambridge University Press, 2003.

Porter, Theodore. *Trust in Numbers.* Princeton, NJ: Princeton University Press, 1996.

Power, Michael. *The Audit Society: Rituals of Verification.* Oxford: Oxford University Press, 1999.

Prado, C. G. *Searle and Foucault on Truth.* New York: Cambridge University Press, 2006.

Prinz, Florian, Thomas Schlange, and Khusru Asadullah. "Believe It or Not: How Much Can We Rely on Published Data on Potential Drug Targets?" *Nature Reviews Drug Discovery* 10, no. 9 (September 2011): 712.

Privitera, Walter. *Problems of Style: Michel Foucault's Epistemology*. Albany: State University of New York Press, 1995.

Proctor, Robert. *Golden Holocaust*. Berkeley: University of California Press, 2012.

Proctor, Robert, and Londa L. Schiebinger. *Agnotology: The Making and Unmaking of Ignorance*. Stanford, CA: Stanford University Press, 2008.

Quine, W. V. "Two Dogmas of Empiricism." In *From a Logical Point of View: Nine Logico-Philosophical Essays*, 2nd revised edition, 20–47. Cambridge, MA: Harvard University Press, 1980.

Rabinow, Paul. *Anthropos Today: Reflections on Modern Equipment*. Princeton, NJ: Princeton University Press, 2003.

"Artificiality and Enlightenment: From Sociobiology to Biosociality." In *Essays on the Anthropology of Reason*, 91–111. Princeton, NJ: Princeton University Press, 1996.

French DNA: Trouble in Purgatory. Chicago: University of Chicago Press, 1999.

Making PCR: A Story of Biotechnology. Chicago: University of Chicago Press, 1996.

"Severing the Ties: Fragmentation and Dignity in Late Modernity." In *Essays on the Anthropology of Reason*, 129–52. Princeton, NJ: Princeton University Press, 1996.

Rabinow, Paul, and Gaymon Bennett. *Designing Human Practices: An Experiment with Synthetic Biology*. Chicago: University of Chicago Press, 2012.

Rasimus, Tuomas, Troels Engberg-Pedersen, and Ismo Dunderberg. *Stoicism in Early Christianity*. Peabody, MA: Hendrickson Publishers, 2010.

Rayner, Timothy. *Foucault's Heidegger: Philosophy and Transformative Experience*. New York: Continuum, 2007.

Reydams-Schils, Gretchen. "Authority and Agency in Stoicism." *Greek, Roman, and Byzantine Studies* 51, no. 2 (May 28, 2011): 296–322.

"Review of Reading Seneca: Stoic Philosophy at Rome by Brad Inwood." *Phoenix* 61, nos. 1–2 (April 1, 2007): 186–9. https://doi.org/10.2307/20304648.

The Roman Stoics: Self, Responsibility, and Affection. Chicago: University of Chicago Press, 2005.

Robbins, Joel. *Becoming Sinners: Christianity and Moral Torment in a Papua New Guinea Society*. Berkeley: University of California Press, 2004.

Rorty, Richard. *Achieving Our Country: Leftist Thought in Twentieth-Century America*. Cambridge, MA: Harvard University Press, 1998.

Rowe, C. Kavin. *One True Life: The Stoics and Early Christians as Rival Traditions*. New Haven, CT: Yale University Press, 2016.

Sarewitz, Daniel. "Saving Science." *New Atlantis*, no. 49 (Spring/Summer 2016): 4–40.

Scherz, China. *Having People, Having Heart*. Chicago: University of Chicago Press, 2014.

Scherz, Paul. "Fragments of the Body in Christian, Bioethical and Social Imaginaries." *Studies in Christian Ethics* 30, no. 4 (November 1, 2017): 449–63.

"Laudato Si' and the Use of Scientific Research in Theology and Public Policy." *The Heythrop Journal* 59, no. 6 (November 2018): 1049–59.

"The Legal Suppression of Scientific Data and the Christian Virtue of Parrhesia." *Journal of the Society of Christian Ethics* 35, no. 2 (2015): 275–301.

Schlier, Heinrich. "παρρησία, παρρησιάζομαι," ed. Gerhard Kittel and Gerhard Friedrich. *Theological Dictionary of the New Testament*. Ann Arbor: Eerdmans, 1967.

Schofer, Jonathan Wyn. *The Making of a Sage: A Study in Rabbinic Ethics*. Madison: University of Wisconsin Press, 2005.

Schuld, J. Joyce. *Foucault and Augustine: Reconsidering Power and Love*. Notre Dame, IN: University of Notre Dame Press, 2003.

Schumacher, Lydia. *Divine Illumination: The History and Future of Augustine's Theory of Knowledge*. Malden, MA: Wiley-Blackwell, 2011.

Sedgwick, Peter H. *The Market Economy and Christian Ethics*. Cambridge: Cambridge University Press, 1999.

Sellars, John. *The Art of Living: The Stoics on the Nature and Function of Philosophy*. Burlington, VT: Ashgate, 2003.

Seneca. *Epistles 93–124*, trans. Richard M. Gummere. Cambridge, MA: Loeb Classical Library, 1925.

Natural Questions, trans. Harry M Hine. Chicago: University of Chicago Press, 2010.

Seneca: Moral Essays, vol. III: *De Beneficiis*, trans. John W. Basore. Cambridge, MA: Loeb Classical Library, 1935.

Sertillanges, A. G. *The Intellectual Life: Its Spirit, Conditions, Methods*. Westminster, MD: Newman Press, 1959.

Servick, Kelly. "Texas Has Sanctioned Unapproved Stem Cell Therapies. Will It Change Anything?" *Science News*, June 15, 2017, www.sciencemag.org/news/2017/06/texas-has-sanctioned-unapproved-stem-cell-therapies-will-it-change-anything.

Shapin, Steven. *The Scientific Life: A Moral History of a Late Modern Vocation*. Chicago: University of Chicago Press, 2008.

Sherkow, Jacob. "Law, History and Lessons in the CRISPR Patent Conflict." *Nature Biotechnology* 33 (2015): 256–7.

Sherwin, Michael. "Infused Virtue and the Effects of Acquired Vice: A Test Case for the Thomistic Theory of Infused Cardinal Virtues." *The Thomist* 73, no. 1 (2009): 29–52.

Sipp, Douglas. "The Unregulated Commercialization of Stem Cell Treatments: A Global Perspective." *Frontiers of Medicine* 5, no. 4 (December 2011) 348–55.

Sismondo, Sergio. "Ghosts in the Machine: Publication Planning in the Medical Sciences." *Social Studies of Science* 39, no. 2 (April 1, 2009): 171–98.

Sismondo, Sergio, and Mathieu Doucet. "Publication Ethics and the Ghost Management of Medical Publication." *Bioethics* 24, no. 6 (July 2010): 273–83.

Skloot, Rebecca. *The Immortal Life of Henrietta Lacks*. New York: Crown Publishers, 2010.

Slater, Dashka. "The Frog of War." *Mother Jones*, February 2012.

Smith, Robin, Tomasz Trafny, and Max Gomez. *The Healing Cell: How the Greatest Revolution in Medical History Is Changing Your Life*. New York: Center Street, 2013.

"Smoke and Mirrors." *Nature* 496, no. 7445 (April 18, 2013): 269.

Sorabji, Richard. *Emotion and Peace of Mind: From Stoic Agitation to Christian Temptation*. New York: Oxford University Press, 2000.

Spaemann, Robert. *Persons: The Difference between "Someone" and "Something,"* trans. Oliver O'Donovan. New York: Oxford University Press, 2017.

Spanneut, Michel. *Le Stoícisme des Pères de L'église, de Clément de Rome à Clément d'Alexandrie*, Patristica Sorbonensia 1 (Paris: Éditions du Seuil, 1957).

Stern, Nicholas. "Building on Success and Learning from Experience: An Independent Review of the Research Excellence Framework." London: Department for Business, Energy & Industrial Strategy, 2016.

"Steven Pinker on Francis Collins." Why Evolution Is True. Accessed August 20, 2013. http://whyevolutionistrue.wordpress.com/2009/07/11/steven-pinker-on-francis-collins/.

Stevenson, Abigail. "World's Youngest Female Billionaire – Next Steve Jobs?" CNBC, September 23, 2015, www.cnbc.com/2015/09/23/worlds-youngest-female-billionaire-next-steve-jobs.html.

Stout, Jeffrey. *Blessed Are the Organized: Grassroots Democracy in America*. Princeton, NJ: Princeton University Press, 2010.

Ethics after Babel: The Languages of Morals and Their Discontents. Boston: Beacon Press, 1988.

Strathern, Marilyn. *Audit Cultures: Anthropological Studies in Accountability, Ethics and the Academy*. New York: Routledge, 2000.

Striker, Gisella. "The Role of Oikeiosis in Stoic Ethics." *Oxford Studies in Ancient Philosophy* 1 (1983): 145–67.

Sullivan-Dunbar, Sandra. *Human Dependency and Christian Ethics*. New York: Cambridge University Press, 2017.

Szakolczai, Árpád. *Max Weber and Michel Foucault: Parallel Life-Works*. New York: Routledge, 1998.

Tachibana, Masahito, Paul Amato, Michelle Sparman, et al. "Human Embryonic Stem Cells Derived by Somatic Cell Nuclear Transfer." *Cell* 153, no. 6 (June 6, 2013): 1228–38.

Takahashi, Kazutoshi, Koji Tanabe, Mari Ohnuki, et al. "Induction of Pluripotent Stem Cells from Adult Human Fibroblasts by Defined Factors." *Cell* 131, no. 5 (November 30, 2007): 861–72.

Tanner, Kathryn. "Grace without Nature," in *Without Nature?*, ed. David Albertson and Cabell King. New York: Fordham University Press, 2010, 363–76.

Taylor, Charles. "Foucault on Freedom and Truth." In *Foucault: A Critical Reader*, ed. David Hoy, 69–102. New York: Basil Blackwell, 1986.

Sources of the Self: The Making of the Modern Identity. Cambridge: Cambridge University Press, 1992.

Thom, Johan Carl. *Cleanthes'* Hymn to Zeus*: Text, Translation, and Commentary*. Tubingen: Mohr Siebeck, 2005.

Thorsteinsson, Runar. *Roman Christianity and Roman Stoicism: A Comparative Study of Ancient Morality*. Oxford: Oxford University Press, 2010.

Tran, Jonathan. *Foucault and Theology*. New York: Bloomsbury T&T Clark, 2011.

Turkle, Sherry. *Alone Together: Why We Expect More from Technology and Less from Each Other*. New York: Basic Books, 2011.

Turner, Leigh, and Paul Knoepfler. "Selling Stem Cells in the USA: Assessing the Direct-to-Consumer Industry." *Cell Stem Cell* 19, no. 2 (August 4, 2016): 154–7.

United States Conference of Catholic Bishops. "On Embryonic Stem Cell Research." United States Conference of Catholic Bishops, 2008.

Unnik, W. C. van. "The Christian's Freedom of Speech in the New Testament." *Bulletin of the John Rylands Library* 44 (1962): 466–88.

Venter, J. Craig. *Life at the Speed of Light: From the Double Helix to the Dawn of Digital Life*. New York: Penguin Books, 2013.

Verhey, Allen. *The Christian Art of Dying: Learning from Jesus*. Grand Rapids, MI: Eerdmans, 2011.

Veyne, Paul. *Seneca: The Life of a Stoic*. New York: Routledge, 2002.

Vogt, Christopher P. *Patience, Compassion, Hope, and the Christian Art of Dying Well*. Lanham, MD: Rowman & Littlefield, 2004.

Vogt, Katja Maria. *Law, Reason, and the Cosmic City: Political Philosophy in the Early Stoa*. New York: Oxford University Press, 2008.

Waldby, Cathy, and Robert Mitchell. *Tissue Economies: Blood, Organs, and Cell Lines in Late Capitalism*. Durham, NC: Duke University Press, 2006.

Walzer, Michael. "The Politics of Michel Foucault." In *Foucault: A Critical Reader*, ed. David Hoy, 51–68. New York: Basil Blackwell, 1986.

Watson, James D. *The Double Helix: A Personal Account of the Discovery of the Structure of DNA*. New York: Atheneum, 1968.

Weber, Max. "Science as a Vocation." In *From Max Weber*, ed. H. H. Gerth and C. Wright Mills, 129–59. New York: Oxford University Press, 1958.

Weinberg, Alvin M. "Science and Trans-Science." *Minerva* 10 (1972): 209–22.

Wellmon, Chad. "Whatever Happened to General Education?" *Hedgehog Review* 19, no. 1 (2017).

White, Lynn. "The Historical Roots of Our Ecologic Crisis." In *Philosophy and Technology*, ed. Carl Mitcham and Robert Mackey, 259–65. New York: Free Press, 1983.

Whyte, William H. *The Organization Man*. Philadelphia: University of Pennsylvania Press, 2002.

Williams, Gareth D. *The Cosmic Viewpoint: A Study of Seneca's Natural Questions*. New York: Oxford University Press, 2012.

Williams, Rowan. "Sapientia and the Trinity: Reflections on De Trinitate." In *Collectanea Augustiniana: Melanges T. J. van Bavel*, ed. Bernard Bruning, Mathijs Lamberigts, and J. van Houtem, 317–32. Louvain: Leuven University Press, 1990.

Wilson, Chris. "Jesus Goes to Bethesda." *Slate*, July 9, 2009. www.slate.com/articles/health_and_science/science/2009/07/jesus_goes_to_bethesda.html.

Wilson, Sloan. *The Man in the Gray Flannel Suit*, 4th edition. New York: Da Capo Press, 2002.

Woolf, S. H. "The Meaning of Translational Research and Why It Matters." *JAMA* 299, no. 2 (January 9, 2008): 211–13. https://doi.org/10.1001/jama.2007.26.

"The World's Billionaires: #435 Elizabeth Holmes." *Forbes Online*, August, 22, 2016. www.forbes.com/profile/elizabeth-holmes/.

Wynne, Brian. "Seasick on the Third Wave? Subverting the Hegemony of Propositionalism." *Social Studies of Science* 33, no. 3 (2003): 401–17.

Yu, Junying, Maxim A. Vodyanik, Kim Smuga-Otto, Jessica Antosiewicz-Bourget, Jennifer L. Frane, Shulan Tian, Jeff Nie, et al. "Induced Pluripotent Stem Cell Lines Derived from Human Somatic Cells." *Science* 318, no. 5858 (2007): 1917–20.

Zahm, John. *Evolution and Dogma*. Chicago: D. H. McBride, 1896.

Zaloom, Caitlin. *Out of the Pits: Traders and Technology from Chicago to London*. Chicago: University of Chicago Press, 2010.

Zhang, Sarah. "How the CRISPR Patent Dispute Became So Heated." *The Atlantic*, December 6, 2016, www.theatlantic.com/science/archive/2016/12/crispr-patent-in-court/509579/.

Zigon, Jarrett. *"HIV Is God's Blessing": Rehabilitating Morality in Neoliberal Russia*. Berkeley: University of California Press, 2010.

Index

TITLES PUBLISHED IN THE SERIES (*continued from page iii*)

14. *The Market Economy and Christian Ethics*, Peter Sedgwick
15. *Churchgoing and Christian Ethics*, Robin Gill
16. *Inequality and Christian Ethics*, Douglas Hicks
17. *Biology and Christian Ethics*, Stephen Clark
18. *Altruism and Christian Ethics*, Colin Grant
19. *The Public Forum and Christian Ethics*, Robert Gascoigne
20. *Evil and Christian Ethics*, Gordon Graham
21. *Living Together and Christian Ethics*, Adrian Thatcher
22. *The Common Good and Christian Ethics*, David Hollenbach
23. *Self-Love and Christian Ethics*, Darlene Fozard Weaver
24. *Economic Compulsion and Christian Ethics*, Albino Barrera
25. *Genetics and Christian Ethics*, Celia Deane-Drummond
26. *Health Care and Christian Ethics*, Robin Gill
27. *Alcohol, Addiction and Christian Ethics*, Christopher C. H. Cook
28. *Human Evolution and Christian Ethics*, Stephen J. Pope
29. *Media Violence and Christian Ethics*, Jolyon Mitchell
30. *Forgiveness and Christian Ethics*, Anthony Bash
31. *Market Complicity and Christian Ethics*, Albino Barrera
32. *Global Justice, Christology and Christian Ethics*, Lisa Sowle Cahill
33. *Lying and Christian Ethics*, Christopher O. Tollefsen
34. *Moral Passion and Christian Ethics*, Robin Gill
35. *Hope and Christian Ethics*, David Elliot
36. *Human Dependency and Christian Ethics*, Sandra Sullivan-Dunbar
37. *Biotechnology, Human Nature, and Christian Ethics*, Gerald McKenny